SpringerBriefs in Agriculture

More information about this series at http://www.springer.com/series/10183

Fernando Ramírez • Jose Kallarackal

Tree Pollination Under Global Climate Change

 Springer

Fernando Ramírez
Independent Researcher
Bogotá, Cundinamarca, Colombia

Jose Kallarackal
Kerala Forest Research Institute
Peechi, Thrissur, Kerala, India

ISSN 2211-808X   ISSN 2211-8098   (electronic)
SpringerBriefs in Agriculture
ISBN 978-3-319-73968-7   ISBN 978-3-319-73969-4   (eBook)
https://doi.org/10.1007/978-3-319-73969-4

Library of Congress Control Number: 2017964227

Printed on acid-free paper

This Springer imprint is published by Springer Nature
The registered company is Springer International Publishing AG
The registered company address is: Gewerbestrasse 11, 6330 Cham, Switzerland

Abstract

Tree pollination has been reported to be impacted by climate change in numerous countries worldwide. Tree pollination is an important ecosystem supporting service that is threatened by global warming and climate change. Warm climatic conditions cause disturbances in pollen physiology. For example, warm temperatures cause decrease in pollen germination and fertilization. Furthermore, insect pollinators are displaced or their life cycles are disturbed by warm conditions leading to great economic losses for fruit tree growers. Whether pollinated by vertebrates, insects, or wind, global warming will have impacts on all these different types of pollination. This book reviews pollination aspects of both wild and cultivated fruit tree species in a global climate change context. It explores cross-pollination mediated by insects and vertebrates, abiotic factors, self-pollination, and their global warming implications. We identify the link between abiotic factors such as precipitation and severe droughts in the context of tree pollination and climate change especially in the tropics. Furthermore, pollination and conservation implications in agriculture as well as wild tree populations are explored. Emphasis has been given to fruit trees growing in tropical, subtropical, and temperate environments.

Preface

It is well known that pollination is an important ecosystem service that is not only essential for plant fertility but also for increasing their productivity. Majority of the plants are pollinated by an interaction between an insect vector and the plant, notwithstanding the importance of abiotic agents such as wind and water. It has been known for many centuries that variation in climatic conditions, which occur in the normal course, has immense effect in determining productivity of many horticultural crops where pollination is essential. However, the loss in productivity suffered during one year was usually compensated by suitable weather conditions in the following year.

Lately, the situation has changed drastically in such a way that the phenomenon of global warming and consequent climate change are affecting the phenological processes in both plants and animals, especially in insects which are the major pollinators. Researches in temperate countries have put an advancement of two weeks for the phenology of plants, especially flowering and anthesis. However, this may not be of any consequence if the pollinator life cycle also synchronizes with the plants. But this is not always happening as expected. Due to increase in temperature, many insect species are not able to complete their life cycle to develop into the adult stage.

Another serious problem is the disruption of the physiology of the pollen-stigma interaction where the pollen grains germinate. Failure of pollen germination and loss of stigma receptivity have been reported in many plants around the world. Sterility has been also reported in several horticultural crops due to lack of fertilization in high temperatures. As the temperatures are predicted to increase further, we can expect more aberrations in both the plant and animal world due to climate change.

In this SpringerBrief, we are trying to review the impact of climate change on pollination biology in the tropics, subtropics, and temperate regions of the world. Much literature has been published in this topic, but not much critical reviewing has been done on the subject. Doing research in two different continents, we find that there is much to be shared in this subject with investigators, students, and farmers around the world. This small book is an attempt to do exactly this. We will be dealing with the impact of precipitation and drought on pollination, which are widely predicted in a climate change situation. The phenological changes happening to the plants and pollinators in a changed climate will be also discussed. The discussion will be centered in terms of both losses in fertility and in productivity. The overall implication to conservation will be also an important point of discussion in this book.

Bogotá, Colombia   Fernando Ramírez
November 26, 2017   Jose Kallarackal

Contents

Chapter 1
Introduction

Pollination is one of the most important ecosystem services. Pollination herein is defined as the movement of pollen grains from the anther to the stigmatic surface of the carpel. It has also been defined as *"the transfer of pollen from the male part of the flower, the anthers, to the receptive female part, the stigma"* (Abrol 2012). Pollination occurs through cross and self-pollination and environmental factors such as animals, water and wind play an important role in this complex process (Fig. 1.1). Pollinators support an estimated 35% of food produced worldwide (Klein et al. 2007). More than 300,000 animal species are considered floral visitors (González-Varo et al. 2013). Furthermore, nearly 75% of important crops worldwide and 80% of the total number of flowering plant species are dependent upon animal pollinators (Nabhan and Buchmann 1997). However, fruit tree pollination is difficult to estimate particularly in tropical environments due to the lack of research studies on this topic. Also, the tropics have a higher diversity of pollinators compared to temperate regions. Increased pollinator diversity augments the pollination service (Brittain et al. 2013). This generates a "subsidiary" effect where multiple pollinators can compensate the lack or reduction of other pollinators. Man-made changes have altered ecosystem habitats and in consequence, have caused negative impacts among animal pollinators, particularly bees (Abrol 2012).

Fruit trees require pollination for fruit development. Pollination in woody angiosperms has been studied in several tropical, temperate and subtropical fruit tree species (Ramírez and Davenport 2013, 2016; Ramírez and Kallarackal 2017) as well as in wild species. They include almond (Klein et al. 2012), apricot (Langridge and Goodman 1981), apple (Ramírez and Davenport 2013), cherry (Holzschuh et al. 2012), carambola (Rodger et al. 2004), coconut (Melendez-Ramirez et al. 2004), litchi (Pandey and Yadava 1970), loquat (Cuevas et al. 2003), mango (Ramírez and Davenport 2016), pear (Monzón et al. 2004), citrus (Davenport 1986), avocado (Vithanage 1990), etc. Fruit crops are an important food and health source for humans worldwide. Fruits contain vitamins, proteins, antioxidants, reducing sugars, fiber, etc., that are known to improve human health and nutrition. Fruits by

© The Author(s) 2018
F. Ramírez, J. Kallarackal, *Tree Pollination Under Global Climate Change*,
SpringerBriefs in Agriculture, https://doi.org/10.1007/978-3-319-73969-4_1

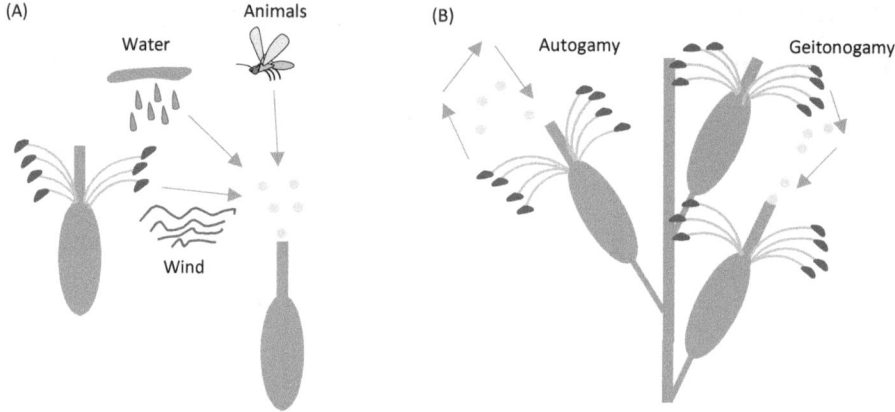

Fig. 1.1 Pollination types. (**a**) Cross-pollination- the transfer of pollen from the stamen to the carpel of plants with different genetic makeup and (**b**) Self-pollination, the transfer of pollen from the stamen to the carpel of the same flower or other flowers within the same plant. Also, occurs between flowers of different plants with identical genetic makeup e.g. some apple cultivars (Illustrations by Fernando Ramírez)

their delicious taste reduce hunger in developed, developing countries and contribute to local and international economies and the wellbeing of people. Pollination is a key feature for fruit production worldwide. Numerous fruit tree species rely on pollination as an effective mechanism to produce fruits and viable seeds contributing to the species perpetuation. However, some fruit tree species have been modified to produce fruits without pollination. This is the case of commercially grown bananas, which are parthenocarpic and produce fruits without the fusion of pollen and egg (Simmonds 1973).

Climate change has been known to impact fruit tree biology in a number of ways, (1) it affects the phenology i.e. advancing or delaying fruiting (Kallarackal and Renuka 2014), (2) elevated CO_2, which affects the process of photosynthesis and growth (Kallarackal and Roby 2012), (3) reduced stomatal conductance and transpiration due to elevated CO_2, (4) affecting the water balance, and (5) affecting the light gathering apparatus (Ramírez and Kallarackal 2015). Climate change impacts whole tree features such as water relations, growth and reproductive aspects such as pollination. The effect of climate change on pollination comprises mechanisms that involve self-, cross-pollination, pollen physiology, floral induction and the influence of environmental factors such as drought, precipitation, temperature, and relative humidity. Furthermore, climate change impacts pollination by altering floral phenology and by affecting the activity of pollinators e.g. flight (Abrol 2012). Higher temperatures generated as a consequence of global warming are responsible for a reduction or increase in phenological cycles in trees (Kallarackal and Renuka 2014; Ramírez and Kallarackal 2015). The advancement of flowering dates has generated advanced flight activity in insect pollinators by 4 days/°C within temperate conditions (Abrol 2012). Also, climate change causes spatial and temporal disparities

Fig. 1.2 Amazonian rainforest sites in Colombia where pollination and climate change studies are required. (**a**) Amacayacu National Park (**b**) tree detail (**c**) insect pollinator, (**d**) Amazon River, (**e**) Loretoyacu River, and (**f**) Amacayacu River (Photos by Fernando Ramírez. Reproduced with permission)

between pollinators and plants (González-Varo et al. 2013). However, the advancement of flowering requires more research studies in the subtropics and tropics. Fig trees in the tropics have been reported to be pollinated by a particular wasp species able to cope with the complex pollination of the fig flowers (Abrol 2012). Thus, the decline of this pollinator due to climate change could have negative effects on fig fruit set. Almond trees in the Himalayas flowered earlier than usual, when no pollinating bees were active, causing a complete crop loss (Abrol 2012). Similarly, climate change has impacted apple pollination. Unusual rains and low temperatures during the flowering season have affected apple tree pollination and fruit set respectively (Abrol 2012). Walnuts and pistachios have floral overlap between male and female flowers that can be affected by insufficient chilling and reduced pollination (Gradziel et al. 2007).

Climate change could affect tropical and temperate regions by unusual rains and warmer than usual temperatures, which are likely to hamper productivity (Wheeler and von Braun 2013). However, warmer climatic conditions could be favorable for subtropical fruit production to some degree, but can also be harmful to several fruits grown in temperate regions (Luedeling 2012). Global climate change could impact pollination physiology and related aspects in both wild and crop trees, however, this requires further research in several environments, e.g. the Amazon (Fig. 1.2). Pollen germination, tube growth, pollen development within anthers and ovule development can be halted by warmer than usual temperatures. In turn, projected global average temperatures in 2100 will average between 1.8 and 4.0 °C higher than the 1980–2000 average (IPCC 2007). This could have a negative impact on pollen physiology of most tree species.

Fig. 1.3 Trees form Cancun, Mexico. More research is required to understand their pollination interactions (Photos by Fernando Ramírez. Reproduced with permission)

Animal pollinators, both arthropods and vertebrates are key vectors in the tree pollination process. Although, commercially grown trees have a less diverse array of pollinators in contrast to trees found in the wild, pollinators have an essential function in fruit production. The majority of crops and wild plants are dependent on pollination mediated by insects, birds, bats and other vertebrates (Abrol 2012). Pollinator declines are well known in temperate plant species (Smith et al. 2015), while less is known from the tropics. Since the 1990s scientists and horticulturists have become aware of pollination declines (Abrol 2012). A number of reasons have been attributed to the global decline in pollinators, namely, climate change, landscape alteration, habitat fragmentation, invasive species, agricultural intensification, pesticide use, spread of diseases, etc. (Kjøhl et al. 2011; González-Varo et al. 2013). A high diversity of fruit trees are grown in the subtropics and tropics and the influence of climate change in pollination has been described for a few species.

Despite the fact that pollination has been widely investigated in relation to crop plants in general, less is known about the interaction of fruit trees and wild tree populations within the context of climate change perturbations (Fig. 1.3). Therefore, the aim of this book is to expand on the current understanding of pollination of fruit trees under climate change conditions. The book focuses on self- cross-pollinations and the influence of environmental factors such as temperature, drought, flooding and precipitation linked to fruit trees in subtropical and tropical environments. It also discusses the implications for conservation and agriculture.

References

Abrol DP (2012) Pollination biology: biodiversity conservation and agricultural production. Springer, New York

Brittain C, Williams N, Kremen C, Klein A-M (2013) Synergistic effects of non-Apis bees and honey bees for pollination services. Proc R Soc B Biol Sci 280:20122767–20122767. https://doi.org/10.1098/rspb.2012.2767

Cuevas J, Hueso JJ, Puertas M (2003) Pollination requirements of loquat (*Eriobotrya japonica* Lindl.), cv. 'Algerie'. Fruits 58:157–165. https://doi.org/10.1051/fruits:2003004

Davenport TL (1986) Avocado flowering. Hortic Rev (Am Soc Hortic Sci) 8:257–289

González-Varo JP, Biesmeijer JC, Bommarco R et al (2013) Combined effects of global change pressures on animal-mediated pollination. Trends Ecol Evol 28:524–530. https://doi.org/10.1016/j.tree.2013.05.008

Gradziel TM, Lampinen B, Connell JH, Viveros M (2007) "Winters" almond: an early-blooming, productive, and high-quality pollenizer for "Nonpareil". Hortscience 42:1725–1727

Holzschuh A, Dudenhöffer J, Tscharntke T (2012) Landscapes with wild bee habitats enhance pollination, fruit set and yield of sweet cherry. Biol Conserv 153:101–107. https://doi.org/10.1016/J.BIOCON.2012.04.032

IPCC (2007) Climate change 2007 synthesis report. Contribution of working groups I, II and III to the fourth assessment report of the intergovernmental panel on climate change [Core writing team, Pachauri RK, Reisinger A (eds)]. IPCC, Geneva, Switzerland, 104 pp

Kallarackal J, Roby TJ (2012) Response of trees to elevated carbon dioxide and climate change. Biodivers Conserv 21:1327–1342

Kallarackal J, Renuka R (2014) Phenological implications for the conservation of Forest trees. In: Kapoor R, Kaur I, Koul M (eds) Plant reproductive biology and conservation. I.K. International, Delhi, pp 90–109

Kjøhl M, Nielsen A, Stenseth N (2011) Potential effects of climate change on crop pollination. FAO, Rome

Klein A-M, Vaissiere BE, Cane JH et al (2007) Importance of pollinators in changing landscapes for world crops. Proc R Soc B Biol Sci 274:303–313. https://doi.org/10.1098/rspb.2006.3721

Klein A-M, Brittain C, Hendrix SD et al (2012) Wild pollination services to California almond rely on semi-natural habitat. J Appl Ecol 49.:no–no. https://doi.org/10.1111/j.1365-2664.2012.02144.x

Langridge D, Goodman R (1981) Honeybee pollination of the apricot cv. Trevatt. Aust J Exp Agric 21:241. https://doi.org/10.1071/EA9810241

Luedeling E (2012) Climate change impacts on winter chill for temperate fruit and nut production: a review. Sci Hortic (Amsterdam) 144:218–229

Melendez-Ramirez V, Parra-Tabla V, Kevan PG et al (2004) Mixed mating strategies and pollination by insects and wind in coconut palm (Cocos nucifera L. (Arecaceae)): importance in production and selection. Agric For Entomol 6:155–163. https://doi.org/10.1111/j.1461-9563.2004.00216.x

Monzón VH, Bosch J, Retana J (2004) Foraging behavior and pollinating effectiveness of Osmia cornuta (Hymenoptera: Megachilidae) and Apis mellifera (Hymenoptera: Apidae) on "Comice" pear. Apidologie 35:575–585. https://doi.org/10.1051/apido:2004055

Nabhan G, Buchmann S (1997) Services provided by pollinators. Nature's Serv Soc Depend Nat Ecosyst so:133–150. https://doi.org/10.1023/a:1023307309124

Pandey RS, Yadava RPS (1970) Pollination of litchi (Litchi chinensis) by insects with special reference to honeybees. J Apic Res 9:103–105. https://doi.org/10.1080/00218839.1970.11100254

Ramírez F, Davenport TL (2013) Apple pollination: a review. Sci Hortic (Amsterdam) 162:188–203

Ramírez F, Davenport TL (2016) Mango (Mangifera indica L.) pollination: a review. Sci Hortic (Amsterdam) 203:158–168

Ramírez F, Kallarackal J (2017) Feijoa [Acca sellowiana (O. Berg) Burret] pollination: a review. Sci Hortic (Amsterdam) 226:333–341. https://doi.org/10.1016/J.SCIENTA.2017.08.054

Ramírez F, Kallarackal J (2015) Responses of fruit trees to global climate change. Springer, SpringerBriefs, New York

Rodger JG, Balkwill K, Gemmill B (2004) African pollination studies: where are the gaps? Int J Trop Insect Sci 24:5–28. https://doi.org/10.1079/IJT20045

Simmonds N (1973) Los plátanos. Blume, Barcelona

Smith MR, Singh GM, Mozaffarian D, Myers SS (2015) Effects of decreases of animal pollinators on human nutrition and global health: a modelling analysis. Lancet 386:1964–1972. https://doi.org/10.1016/S0140-6736(15)61085-6

Vithanage V (1990) The role of the European honeybee (Apis mellifera L.) in avocado pollination. J Hortic Sci 65:81–86. https://doi.org/10.1080/00221589.1990.11516033

Wheeler T, von Braun J (2013) Climate change impacts on global food security. Science 341:508–513. https://doi.org/10.1126/science.1239402

Chapter 2
Cool, Warm Temperatures and Tree Pollination

In the following section, we provide a detailed description of the flower and its parts because it is relevant to the study of pollination (see Fig. 2.1 for floral parts). Flowers are structures, which consist of an array of parts (organs) borne on a central axis called the receptacle (Rudall 2007). The whole floral structure is suspended by the peduncle, a stalk that attaches the flower to the plant (Glimn-Lacy and Kaufman 2006). The flower might be supported by leaf-like structures called bracts, which are absent or present depending on plant species (Rudall 2007). The typical flower is composed of the internal sexual parts, namely organs, which are covered by sepals and petals (Abrol 2012). The floral perianth or the outer structure of the flower is composed of the sepals referred to as the first whorl or calyx and the petals, which comprises the second whorl or corolla (Rudall 2007). Sepals have a protective function during floral development, are green but can also attain color in some plant species (Glimn-Lacy and Kaufman 2006). Petals are colored parts that function as pollinator attractants via color, shape and pattern (Abrol 2012). The male component of the flower is called the stamen which bears the filament and anther. Anthers commonly bear two pollen sacs at the upper end (Glimn-Lacy and Kaufman 2006). Within the anthers, pollen is produced through the process of microsporogenesis. Among woody angiosperms, pollen is generated through meiosis and further maturation occurs in the anthers (Ramírez and Davenport 2010). Once pollen becomes mature the anthers dehisce or split open releasing pollen grains (Abrol 2012). The female part of the flower is called the carpel or pistil and is composed of the upper end, stigma, mid part, style and lower part ovary (Fig. 1.3). Pollen contacts the stigmatic surface during pollination, and then germinates through the style reaching the ovary, which contains the ovules (Glimn-Lacy and Kaufman 2006). The ovule contains the megaspores; one of these develops into an embryo sac containing an egg (Glimn-Lacy and Kaufman 2006).

Flowers among woody angiosperm species vary in the number of stamens (male parts) and carpels (female parts). Here we refer to the flowers of fruit trees as solitary

© The Author(s) 2018
F. Ramírez, J. Kallarackal, *Tree Pollination Under Global Climate Change*,
SpringerBriefs in Agriculture, https://doi.org/10.1007/978-3-319-73969-4_2

Fig. 2.1 Floral parts and flowers in (**a**) feijoa (*Acca sellowiana*), (**b**) Lulo (*Solanum quitoense* L.), (**c**) capuli cherry (*Prunus serotina* var. capuli), (**d**) guava (*Psidium guajava* L.), (**e**) avocado (*Persea americana* L.), (**f**) "Valencia" orange (*Citrus sinensis*) and (**g**) "Anna" apple (*Malus domestica*) (Photos and illustrations by Fernando Ramírez. Reproduced with permission)

or grouped within inflorescences. The flowers can be staminate (male) carpellate (female) or hermaphrodite, e.g. mango (Fig. 2.2).

Pollen grains require specific environmental conditions for subsequent germination and growth, which is species and cultivar specific. For example, 'Kent' mango (*Mangifera indica* L.) trees require an optimum temperature of 30 °C for effective pollen germination (Dag et al. 2000). Pollination is highly sensitive to temperature extremes across all species, having a negative impact on production (Hatfield and Prueger 2015; Ramírez and Davenport 2016; Ramírez and Kallarackal 2017). Either high or low temperatures are known to affect the pollination physiology of trees. Cool temperature values below 12 °C during flowering interfered with pollination and/or fertilization in mango (Whiley et al. 1988). Cool temperatures during pollen development cause a reduction in the overall pollen viability (Issarakraisila

A

Fig. 2.2 Mango flowers (**a**) pedicel with staminate flowers. Note the five petals, ovary, stamen, staminoids and floral disk, (**b**) hermaphrodite flower and (**c**) carpellate flower (Photos from a mango tree from Puerto Nariño in the Colombian Amazon. Photos taken by Fernando Ramírez. Reproduced with permisssion after Ramírez and Davenport (2016)

1994; Huang et al. 2010). Cool temperatures <16° C have been known to cause floral deformation, delayed pollen germination, pollen-tube growth and induce ovule abortion leading to production of seedless fruits in mango (Young and Sauls 1979). Likewise, guava trees experience floral abscission in response to extremely cool temperatures (Dinesh and Reddy 2012). Date palm inflorescences exposed to cool temperatures (20 °C day/ 8 °C night) were vulnerable to low pollen germination and delayed pollen germination after 3 days (Slavković et al. 2016). Low temperatures, 10 and 15 °C, caused reduced pollen tube elongation in sugar apple (*Annona squamosa* L.) (Rodrigues et al. 2016).

Cool temperatures as a result of climate change can also impair pollinator activity. Cooler than usual temperatures reduce apple blossoming and reduce pollinating activity of bees (Vedwan and Rhoades 2001). Also bee immobilization is caused by low temperatures caused by late snowfall in the Western Himalayas of India (Vedwan and Rhoades 2001). Some bee species such as the Himalayan honey bee *Apis cerana*, begin foraging at temperature as low as 7 °C, whereas, the introduced honey bee, *Apis mellifera* start to forage only at 13 °C (Vedwan and Rhoades 2001).

Temperate fruit trees such as pear, apple, cherry and plum need an optimum temperature between 20 and 25 °C (Rai et al. 2015). High temperatures above the optimum can lead to ineffective pollination and fertilization due to functional damage within the floral-pollen structures. For example, temperatures above 24 °C reduced pollen tube growth in *Prunus domestica* (DeCeault and Polito 2010). Pollen of wild *Prunus* species failed to germinate at warm temperatures (50 °C) (Sorkheh et al. 2011). *Prunus* species are negatively impacted by high temperatures during pollination particularly during pollen germination, tube growth, ovule development and longevity (Jefferies et al. 1982; Moreno et al. 1992; Cerović et al. 2000). However, Hedhly et al. (2005) reported that high temperatures (20–30 °C) increased pollen germination and pollen (*in vitro*) tube growth, and caused loss of stigmatic receptivity in peach. Also, increasing temperatures up to a certain level hasten pollen tube growth in sweet cherry (Hedhly et al. 2003, 2004). Furthermore, Kozai et al. (2004) reported increasing temperature (30 °C) decreases floral size, and pollen germination was halted in 'Hakuho' peach trees. Furthermore, high temperatures cause ovule degeneration in cherry and plum and stigma degeneration in sweet cherry (Postweiler et al. 1985). Under warm temperatures (Average 30 °C) 'Julieta' apple trees experienced anther loss and a reduced capacity to release pollen grains in Brazil (Monteiro et al. 2015). High temperatures in apple inhibit pollen production, floral induction, and reduce viability (Van Marrewijk 1993). Furthermore, apple pollinator activity such as that of bees is reduced below temperatures of 13 °C (Keogh et al. 2010). On the other hand, during temperature rises following pollination, pollen-tubes grow more rapidly, within limits, but the time during which the ovule is receptive is reduced (Dennis 2003; Ramírez and Davenport 2013). Temperature extremes, either high or low interfere with pollination, fertilization and fruit set in apple (Rai et al. 2015).

The unusually high temperatures as a result of climate change can have adverse effect on chilling hours required by temperate and sub-tropical trees. Temperate fruit tree plantations at the end of the nineteenth century occurred in warmer areas than those of traditionally cultivated locations (Campoy et al. 2011). Apple trees grown in warm winter regions with inadequate chilling requirements cause trees to have a number of problems such as phenology referent to bud break, flowering, growth and development of both fruits and trees (Petrí et al. 2012). Among cross-pollinated fruits e.g. pistachios and walnuts insufficient chilling can affect pollination (Gradziel et al. 2007; Rai et al. 2015). Similarly, insufficient chilling causes reduced pollination in peaches (Rai et al. 2015).

Climate change has been known to cause higher than usual temperatures affecting pollination of fruit crops. Under climate change conditions, fruit trees growing in tropical regions with high temperatures could be negatively impacted (Rajan 2012). High temperatures result in floral abscission in carpellate and hermaphrodite flowers and sex-change in bisexual papaya plants (Dinesh and Reddy 2012). Tree pollination in tropical and subtropical regions could be affected by warm temperatures generated as a result of climate warming. Lychee (*Litchi chinensis*) trees failed to produce flowers when exposed to 20 °C for 8 or more hours daily (Menzel and Simpson 1995). In India, mango trees flowered 3 months earlier than usual in 2007 due to warmer temperatures (Abrol 2012). High temperatures (22 °C day/27 °C and night 32 °C/27 °C) have adverse effects on pollen development in lychee (Stern and Gazit 1998). Likewise, high temperatures (30 °C) caused a decrease in pollen germination in Citrus (Distefano et al. 2012). Moreover, higher than normal temperature (33 °C day/28 °C night) conditions during flowering have been known to cause floral abscission and ovule damage in avocado (Sedgley 1977; Argaman 1983; Davenport 1986). *In vitro* feijoa (*Acca sellowiana*) pollen germination was lower at 30 °C (Franzon et al. 2005; Ramírez and Kallarackal 2017). Pollen germination was halted at 45 °C in sugar apple (Rodrigues et al. 2016). Many tropical tree species flower during a brief reproductive period that is at risk if temperatures are high and consequently impede pollination and further floral development (Goldstein and Santiago 2016). Other tropical fruits compensate this by having two or more reproductive events yearly (Ramírez and Kallarackal 2017, 2018). Furthermore, an increased floral production could compensate the brief flower induction periods.

In wild tropical forest trees, extremely warm, rainy or dry conditions could lead to flowering and pollination inhibition (Corlett and LaFrankie 1998). On the other hand, subtropical trees in the wild, exposed to higher than usual temperatures (32 °C) have been known to reduce their pollen viability (Maiti and Rodriguez 2015). Temperate tree pollen germination has been known to be affected by high temperatures. For example, the narrow leaved ash (*Fraxinus angustifolia* Vahl) pollen germination was significantly reduced at high temperatures (25 °C) (Kremer and Jemrić 2006). In the case of temperate wild trees, in 24 species of birch (*Betula*), pollen shedding date was advanced with increasing temperatures (Miller-Rushing and Primack 2008). More research and modeling studies are required to understand the reproductive physiology of trees in relation to predicted climate change scenarios within tropical, temperate and subtropical environments.

Numerous investigations have given documentary evidences for the advancement of pollen seasons due to their link to allergic reactions of pollen in humans. In this section we provide a brief overview of this. In urban areas, early pollination has been documented in *Alnus, Ulmus, Betula,* and *Corylus* as a result of warmer spring caused by climate change (Emberlin et al. 2007; Jager et al. 1996; Van Vliet et al. 2001). Birch (*Betula*) pollen production and release occurs during April to mid-May, but in the last 30 years, there has been a shift to earlier start (Emberlin et al. 1997). Furthermore, *Betula* pollen start dates in London, Brussels, Zurich and Vienna is expected to advance by about 6 days over the next 10 years (Emberlin et al. 2002). In Switzerland, pollen data from 38 years revealed that flowering in *Betula* occurred 15 days earlier and pollen season also started earlier (Frei and Gassner 2008). Similarly, birch pollen season started about 19 days in advance in 2001 in contrast to the 1980's, as a result of climate change (Clot 2001). Moreover, birch pollen season occurred 5 days earlier in the last decade in contrast to the previous 10 years in three cities, namely, Cardiff, Derby and London (Emberlin et al. 1997). While Newnham et al. (2013) found that the start of birch pollen season is sensitive to temperatures in the UK, no clear trend showing changes in pollen season has been observed. However, early flowering *Alnus* spp. and *Corylus* spp. have shown changes in pollen season during the consecutive years from 1996 to 2005 at

Worcester, UK (Emberlin et al. 2007). In *Alnus* and *Corylus* trees higher than usual temperatures towards the end of winter caused an earlier start in pollen season in SW Poland (Malkiewicz et al. 2016). In Denmark, birch pollen season started 14 and 17 days earlier at two different locations more than 200 km apart due to increasing temperatures (Rasmussen 2002). Pollen seasons as a result of warming have started earlier in birch (*Betula*) and oak (*Quercus*) across the continental United Sates (Zhang et al. 2015). In the USA, birch and oak pollen seasons in 2001–2011 started earlier in contrast to the period between 1994 and 2000 with increasing trends of peak values and annual mean (Zhang et al. 2014). In birch, oak and pine trees, pollen seasons have started earlier in 2013 in contrast to 1973. This phenological change in pollen season is linked to the increasing air temperature in Stockholm (Lind et al. 2016).

Tree species such as the Japanese cedar (*Cryptomeria japonica*), which is a gymnosperm, has advanced its pollen season from mid-March to late February and has increased pollen counts. This has been associated with warmer than usual temperature due to climate change from 1983 through 1998 in Toyama City, Japan (Teranishi et al. 2000). In other species, such as *Quercus*, pollination season could occur months earlier and thus, airborne pollen concentrations would increase by 50% with respect to current levels, and bear higher levels within the Mediterranean mainland (García-Mozo et al. 2006).

The duration of pollen season has also been documented to increase for several tree species in Europe. Between 1981 and 2007, the increase of pollen season in olive has been by 18 days, which correlate with increase in number of days with a temperature greater than 30 °C, direct radiation, and overall temperature (Ariano et al. 2010). Moreover, olive flowering phenology could be considered a sensitive indicator of climate fluctuations and their effects within the Mediterranean area (Galán et al. 2005). Recently García-Mozo et al. (2014) found that olive season is occurring earlier, pollen peak has been occurring faster and its end occurs progressively later, changes which are associated with temperature increase within a 30-year period (1982–2011) at Córdoba, Spain. The rising evidence of pollen season advancements and delays has been investigated thoroughly in temperate environments, but more research is required in tropical environments, where the specific response of trees remains greatly unknown. In the tropics, trees respond differently to climate change conditions in contrast to temperate regions and these responses need to be quantified.

References

Abrol DP (2012) Pollination biology: biodiversity conservation and agricultural production. Springer, New York

Argaman E (1983) Effect of temperature and pollen source on fertilization, fruit set and abscission in avocado (*Persea americana* Mill.). Hebrew University of Jerusalem. (in Hebrew). MS thesis, Hebrew University, Jerusalem, Israel

Ariano R, Canonica GW, Passalacqua G (2010) Possible role of climate changes in variations in pollen seasons and allergic sensitizations during 27 years. Ann Allergy Asthma Immunol 104:215–222. https://doi.org/10.1016/j.anai.2009.12.005

Campoy JA, Ruiz D, Egea J (2011) Dormancy in temperate fruit trees in a global warming context: a review. Sci Hortic (Amsterdam) 130:357–372

Cerović R, Ružić Đ, Mićić N (2000) Viability of plum ovules at different temperatures. Ann Appl Biol 137:53–59. https://doi.org/10.1111/j.1744-7348.2000.tb00056.x

Clot B (2001) Airborne birch pollen in Neuchâtel (Switzerland): onset, peak and daily patterns. Aerobiologia (Bologna) 17:25–29. https://doi.org/10.1023/A:1007652220568

Corlett RT, LaFrankie JV (1998) Potential impacts of climate change on tropical Asian forests through an influence on phenology. Clim Chang 39:439–453. https://doi.org/10.102 3/a:1005328124567

Dag A, Eisenstein D, Gazit S (2000) Effect of temperature regime on pollen and the effective pollination of "Kent" mango in Israel. Sci Hortic (Amsterdam) 86:1–11. https://doi.org/10.1016/S0304-4238(99)00134-X

Davenport TL (1986) Avocado flowering. Hortic Rev (Am Soc Hortic Sci) 8:257–289

DeCeault MT, Polito VS (2010) High temperatures during bloom can inhibit pollen germination and tube growth, and adversely affect fruit set in the *Prunus domestica* cultvars "Improved French" and "Muir Beauty.". Acta Hortic 874:163–168. https://doi.org/10.17660/ActaHortic.2010.874.22

Dennis F (2003) Flowering, pollination and fruit set and development. In: Apples botany production and uses. CABI Publishing, Wallingford, pp 153–166. https://doi.org/10.1079/9780851995922.0000

Dinesh M, Reddy B (2012) Physiological basis of growth and fruit yield characteristics of tropical and sub-tropical fruits to temperature. In: Sthapit B, Ramanatha S, Sthapit R (eds) Tropical fruit tree species and climate change. Bioversity International, New Delhi, pp 45–70

Distefano G, Hedhly A, Las Casas G et al (2012) Male-female interaction and temperature variation affect pollen performance in Citrus. Sci Hortic (Amsterdam) 140:1–7. https://doi.org/10.1016/j.scienta.2012.03.011

Emberlin J, Jones S, Mullins J et al (1997) The trend to earlier birch pollen seasons in the U.K.: a biotic response to changes in weather conditions? Grana 36:29–33. https://doi.org/10.1080/00173139709362586

Emberlin J, Detandt M, Gehrig R et al (2002) Responses in the start of *Betula* (birch) pollen seasons to recent changes in spring temperatures across Europe. Int J Biometeorol 46:159–170. https://doi.org/10.1007/s00484-002-0139-x

Emberlin J, Smith M, Close R, Adams-Groom B (2007) Changes in the pollen seasons of the early flowering trees *Alnus* spp. and *Corylus* spp. in Worcester, United Kingdom, 1996–2005. Int J Biometeorol 51:181–191. https://doi.org/10.1007/s00484-006-0059-2

Franzon R, Corrêa E, Raseira M (2005) In vitro pollen germination of feijoa (*Acca sellowiana* (Berg) Burret). Crop Breed Appl Biotechnol 5:229–233

Frei T, Gassner E (2008) Climate change and its impact on birch pollen quantities and the start of the pollen season an example from Switzerland for the period 1969–2006. Int J Biometeorol 52:667–674. https://doi.org/10.1007/s00484-008-0159-2

Galán C, García-Mozo H, Vázquez L et al (2005) Heat requirement for the onset of the *Olea europaea* L. pollen season in several sites in Andalusia and the effect of the expected future climate change. Int J Biometeorol 49:184–188. https://doi.org/10.1007/s00484-004-0223-5

García-Mozo H, Galán C, Jato V et al (2006) *Quercus* pollen season dynamics in the Iberian Peninsula: response to meteorological parameters and possible consequences of climate change. Ann Agric Environ Med 13:209–224

García-Mozo H, Yaezel L, Oteros J, Galán C (2014) Statistical approach to the analysis of olive long-term pollen season trends in southern Spain. Sci Total Environ 473–474:103–109. https://doi.org/10.1016/j.scitotenv.2013.11.142

Glimn-Lacy J, Kaufman PB (2006) Botany illustrated. Introduction to plants, major groups, flowering plant families. Springer, New York

Goldstein G, Santiago L (2016) Tree physiology tropical tree physiology adaptations and responses in a changing environment. Springer, Cham, p 467. https://doi.org/10.1007/978-3-319-27422-5

Gradziel TM, Lampinen B, Connell JH, Viveros M (2007) "Winters" almond: an early-blooming, productive, and high-quality pollenizer for "Nonpareil". Hortscience 42:1725–1727

Hatfield JL, Prueger JH (2015) Temperature extremes: effect on plant growth and development. Weather Clim Extrem 10:4–10. https://doi.org/10.1016/j.wace.2015.08.001

Hedhly A, Hormaza JI, Herrero M (2003) The effect of temperature on stigmatic receptivity in sweet cherry (*Prunus avium* L.) Plant Cell Environ 26:1673–1680. https://doi.org/10.1046/j.1365-3040.2003.01085.x

Hedhly A, Hormaza JI, Herrero M (2004) Effect of temperature on pollen tube kinetics and dynamics in sweet cherry, *Prunus avium* (Rosaceae). Am J Bot 91:558–564. https://doi.org/10.3732/ajb.91.4.558

Hedhly A, Hormaza JI, Herrero M (2005) The effect of temperature on pollen germination, pollen tube growth, and stigmatic receptivity in peach. Plant Biol 7:476–483. https://doi.org/10.1055/s-2005-865850

Huang J-H, Ma W-H, Liang G-L et al (2010) Effects of low temperatures on sexual reproduction of "Tainong 1" mango (*Mangifera indica*). Sci Hortic (Amsterdam) 126:109–119. https://doi.org/10.1016/j.scienta.2010.06.017

Issarakraisila M (1994) Effects of temperature on pollen viability in mango cv. "Kensington". Ann Bot 73:231–240

Jager S, Nilsson S, Berggren B et al (1996) Trends of some airborne tree pollen in the Nordic countries and Austria, 1980–1993 a comparison between Stockholm, Trondheim, Turku and Vienna. Grana 35:171–178

Jefferies C, Brain P, Stott K, Belcher A (1982) Experimental systems and a mathematical model for studying temperature effects on pollentube growth and fertilization in plum. Plant Cell Environ 5:231–236. https://doi.org/10.1111/1365-3040.ep11572417

Keogh R, Robinson A, Mullins I (2010) Pollination awareness case study: apple. Rural Industries Research and Development Corporation, Leederville

Kozai N, Beppu K, Mochioka R et al (2004) Adverse effects of high temperature on the development of reproductive organs in "Hakuho" peach trees. J Hortic Sci Biotechnol 79:533–537. https://doi.org/10.1080/14620316.2004.11511801

Kremer D, Jemrić T (2006) Pollen germination and pollen tube growth in *Fraxinus pennsylvanica*. Biologia (Bratisl) 61:79–83. https://doi.org/10.2478/s11756-006-0011-2

Lind T, Ekebom A, Kübler KA et al (2016) Pollen season trends (1973-2013) in Stockholm area, Sweden. PLoS One 11:e0166887. https://doi.org/10.1371/journal.pone.0166887

Maiti R, Rodriguez HG (2015) Phenology, morphology and variability in pollen viability of four woody species (*Cordia boissieri, Parkinsonia texana, Parkinsonia aculeate* and *Leucophyllum frutescens*) exposed to environmental temperature in Linares, northeast of Mexico. For Res Open Access 4:002–S1: 002. https://doi.org/10.4172/2168-9776.S1-002

Malkiewicz M, Drzeniecka-Osiadacz A, Krynicka J (2016) The dynamics of the *Corylus, Alnus,* and *Betula* pollen seasons in the context of climate change (SW Poland). Sci Total Environ 573:740–750. https://doi.org/10.1016/j.scitotenv.2016.08.103

Menzel CM, Simpson DR (1995) Temperatures above 20°C reduce flowering in lychee (*Litchi chinensis* Sonn.) J Hortic Sci 70:981–987. https://doi.org/10.1080/14620316.1995.11515374

Miller-Rushing AJ, Primack RB (2008) Effects of winter temperatures on two birch (*Betula*) species. Tree Physiol 28:659–664. https://doi.org/10.1093/treephys/28.4.659

Monteiro VM, da Silva CI, de SP FAJ, Freitas BM (2015) Floral biology and implications for apple pollination in semiarid northeastern Brazil. J Agric Environ Sci 4:42–50. https://doi.org/10.15640/jaes.v4n1a6

Moreno Y, Miller-Azarenko A, Potts W (1992) Genotype, temperature, and fall-applied Ethephon affect plum flower bud development and ovule longevity. J Am Soc Hortic Sci 117:14–21

Newnham RM, Sparks TH, Skjøth CA et al (2013) Pollen season and climate: is the timing of birch pollen release in the UK approaching its limit? Int J Biometeorol 57:391–400. https://doi.org/10.1007/s00484-012-0563-5

Petrí J, Hawerroth F, Leite G et al (2012) Apple phenology in subtropical climate conditions. In: Zhang X (ed) Phenology and climate change. InTech, Rijeka, pp 195–216

Postweiler K, Stösser R, Anvari SF (1985) The effect of different temperatures on the viability of ovules in cherries. Sci Hortic (Amsterdam) 25:235–239. https://doi.org/10.1016/0304-4238(85)90120-7

Rai R, Joshi S, Roy S et al (2015) Implications of changing climate on productivity of temperate fruit crops with special reference to apple. J Hortic. https://doi.org/10.4172/2376-0354.1000135

Rajan S (2012) Phenological responses to temperature and rainfall: a case study of mango. In: Sthapit BR, Ramanatha RV, Sthapit S (eds) Tropical fruit tree species and climate change. Biodiversity International, New Delhi, pp 71–96

Ramírez F, Davenport TL (2010) Mango (*Mangifera indica* L.) flowering physiology. Sci Hortic (Amsterdam) 126:65–72. https://doi.org/10.1016/j.scienta.2010.06.024

Ramírez F, Davenport TL (2013) Apple pollination: a review. Sci Hortic (Amsterdam) 162:188–203

Ramírez F, Davenport TL (2016) Mango (*Mangifera indica* L.) pollination: a review. Sci Hortic (Amsterdam) 203:158–168

Ramírez F, Kallarackal J (2017) Feijoa [*Acca sellowiana* (O. Berg) Burret] pollination: a review. Sci Hortic (Amsterdam) 226:333–341. https://doi.org/10.1016/J.SCIENTA.2017.08.054

Ramírez F, Kallarackal J (2018) Phenological growth stages of Feijoa [*Acca sellowiana* (O. Berg) Burret] according to the BBCH scale under tropical Andean conditions. Sci Hortic (Amsterdam) 232:184–190

Rasmussen A (2002) The effects of climate change on the birch pollen season in Denmark. Aerobiologia (Bologna) 18:253–265. https://doi.org/10.1023/A:1021321615254

Rodrigues BRA, dos Santos RC, Nietsche S et al (2016) Determinação de temperaturas cardinais em pinheira (*Annona squamosa* L.) Cienc e Agrotecnologia 40:145–154. https://doi.org/10.1590/1413-70542016402039115

Rudall P (2007) Anatomy of flowering plants: an introduction to structure and development. Cambridge University Press, Cambridge

Sedgley M (1977) The effect of temperature on floral behaviour, pollen tube growth and fruit set in the avocado. J Hortic Sci 52:135–141

Slavković F, Greenberg A, Sadowsky A et al (2016) Effects of applying variable temperature conditions around inflorescences on fertilization and fruit set in date palms. Sci Hortic (Amsterdam) 202:83–90. https://doi.org/10.1016/j.scienta.2016.02.030

Sorkheh K, Shiran B, Rouhi V, Khodambashi M (2011) Influence of temperature on the in vitro pollen germination and pollen tube growth of various native Iranian almonds (*Prunus* L. spp.) species. Trees 25:809–822. https://doi.org/10.1007/s00468-011-0557-7

Stern RA, Gazit S (1998) Pollen viability in lychee. J Am Soc Hortic Sci 123:41–46

Teranishi H, Kenda Y, Katoh T et al (2000) Possible role of climate change in the pollen scatter of Japanese cedar *Cryptomeria japonica* in Japan. Clim Res 14:65–70

Van Marrewijk G (1993) Flowering biology and hybrid varieties. Hybrid varieties.International Course on Applied Plant Breeding. The Netherlands, IAC

Van Vliet A, Den Dulk J, De Groot R (2001) The times they are a changing. Environ Syst Anal Group Wageningen Univ, Wageningen

Vedwan N, Rhoades RE (2001) Climate change in the Western Himalayas of India: a study of local perception and response. Clim Res 19:109–117. https://doi.org/10.3354/cr019109

Whiley A, Saranah J, Rasmussen T, et al (1988) Effect of temperature on ten mango cultivars with relevance to production in Australia. In: Proceedings of the fourth Australasian conference on tree and nut crops. Lismore, New South Wales, pp 176–185

Young T, Sauls J (1979) The mango industry in Florida. University of Florida, Cooperative Extension Service. Bulletin, 189, Gainesville

Zhang Y, Bielory L, Georgopoulos PG (2014) Climate change effect on *Betula* (birch) and *Quercus* (oak) pollen seasons in the United States. Int J Biometeorol 58:909–919. https://doi.org/10.1007/s00484-013-0674-7

Zhang Y, Bielory L, Mi Z et al (2015) Allergenic pollen season variations in the past two decades under changing climate in the United States. Glob Chang Biol 21:1581–1589. https://doi.org/10.1111/gcb.12755

Chapter 3
Precipitation, Flooding and Pollination

Climate change has altered the rainfall patterns worldwide. In the last century, the amount of annual precipitation and occurrence of extreme precipitation events have increased worldwide (Rosenzweig et al. 1996). Under climate change conditions, unseasonal rains often occur during dry periods, or as extended rainy seasons causing flooding events. Magrin et al. (2014) reported an increase in climate change driven extreme events i.e. flooding, droughts, heavy rains, landslides, heat waves in Central and South America (Fig. 3.1). The rainfall pattern drives climate regulation within a biome (Scarano and Ceotto 2015). It is also a key to controlling watershed levels and soil stability on mountain slopes (Scarano and Ceotto 2015).

The unpredictable pattern of climate change extreme events such as hurricanes, tornadoes, etc., have caused severe flooding and devastating events. Recently, hurricanes such as Irma, and Maria, have caused flooding and devastation in a number of Caribbean Islands, e.g. Puerto Rico, US Virgin Islands, etc. and coastal areas within the US mainland such as Florida. These events have torn down many trees and it is likely that pollinators could have been highly impacted. Bee species richness declined (with a loss of 40% species occurring before) after Hurricane Dean that impacted the Yucatan Peninsula, Mexico (Ramírez et al. 2016). This event also reduced the number of social, parasocial, solitary bee species, and the ones nesting within cavities and wood (Ramírez et al. 2016). In 1992, hurricane Andrew impacted Florida, USA, and had devastating effects on trees by completely uprooting trees, braking branches, detaching leaves, fruits and flowers. Trees like grapefruit lost all fruits and in some cases they were torn, or completely uprooted at Cooper City, Broward County, Florida, U.S.A. (Ramírez pers. Obs.). Hurricane Andrew had devastating effects on fruit trees, which were toppled, stumped, destroyed or left standing with only major limbs (Crane et al. 1993). These authors conducted several surveys within orchards determining the degree of impact 10–15 months after the hurricane, finding that trees within orchards such as lime (95%), carambola (93%), antemoya (90%), avocado (87%), survived the hurricane, whereas, mango (71%), longan (70%) and lychee (60%) had a greater hurricane impact (Crane et al. 1993).

© The Author(s) 2018
F. Ramírez, J. Kallarackal, *Tree Pollination Under Global Climate Change*,
SpringerBriefs in Agriculture, https://doi.org/10.1007/978-3-319-73969-4_3

Fig. 3.1 The flooded forest in the Amazon River near Leticia, Colombia. Floods in the Amazon have been higher that usual over the last years and dry periods longer (Photos by Fernando Ramírez. Reproduced with permission)

However, little is known about the effect of hurricane Andrew on fruit tree pollinators. Trees such as *Ficus aurea* damaged by hurricane Andrew comprised branch, leaf and fruit loss, as well as the presumed local extinction of its pollinator, the fig wasp *Pegoscapus jimenezi* (Bronstein and Hossaert-McKey 1995). After 5 months, post-hurricane, the fig wasp abundance and the fig flowering phenology showed a recovery reaching near pre-hurricane levels (Bronstein and Hossaert-McKey 1995). The pollination of the shrub Bahama Swamp-bush (*Pavonia bahamensis*) was limited by the absence of its bird pollinators, Bananaquits and Bahama Woodstars, which were rarely seen at San Salvador Island during 1996–1997 due to the impact of Hurricane Lili (Category 2) (Rathcke 2000). Fruit bats, *Pteropus samoensis* and *P. tonganus*, have been considered important tree pollinators that historically showed population declines of 80–90% due to Hurricane Ofa that impacted the American Samoa Islands in the 1990, aside from hunting (Craig et al. 1994). Hurricanes could also negatively impact the reproductive success of white mangrove tree (*Laguncularia racemosa*), which is considered to be pollinated by insects and self-compatible (Landry 2013). This is due to changing the insect pollinator species abundances and community composition, e.g. after two hurricanes within the Florida area, species richness values were reduced and ranged between 43% and 65% and diversity declined by 36–70% (Landry 2013). Pollination services are greatly disturbed by the effect of hurricanes. More investigations are needed to examine the effect of hurricanes on tree pollinators in detail.

Individual tree responses to climate change will depend on their dispersal abilities, physiology, life-history strategies, and dispersal abilities that can mediate the responses to potential threats (Potter et al. 2017). Also, reproductive physiology aspects such as pollination, pollen dispersal, pollen-stigma interaction and the subsequent steps leading to fertilization, fruit development and fruit set could be affected by climate change driven extreme events such as heavy rainfall (Snyder 2017). In the tropics, temperature and incoming solar radiation are relatively uniform throughout the year, while, there are considerable fluctuations in rainfall (Chapman et al. 2005; Corlett and LaFrankie 1998). Within these conditions, tree pollen grain release and dispersal can be affected by heavy rainfall. For example, heavily overcast and rainy conditions delay or halt mango flowering in the tropics (Ramírez and Davenport 2012) and also have a negative impact on pollination. However, more research is called for to examine the effects of precipitation on pollen dispersal under climate change conditions.

Under unusual precipitation conditions, pollinators reduce their activity. For example, late snow events as a result of climate change affect the pollination process by causing bee immobilization due to low temperatures and producing temporal mismatches in the Himalayas (Negi et al. 2017). Furthermore, climate change conditions such as prolonged cloudy overcast and rainy days and rains during full flowering hinder cross-pollination and fruit set in litchi (Kumar 2014). Also, during flowering, precipitation could decrease pollination and fruit set in citrus under climate change conditions (Rosenzweig et al. 1996). Unusual rains in mango orchards at La Mesa, Colombia, cause insect pollinators such as wild and native bees to halt their activity and remain within their hives until more favorable weather conditions prevail. Also, trees become highly asynchronous, due to unusual overcast and rainy conditions and few sunny intervals, generating less flowering (Ramírez and Kallarackal 2015). Less flowering and reduced pollinator activity can result in pollinator mismatches and less fruit production, which have occurred over the 2011–2013 period (Ramírez and Kallarackal 2015). This could hamper the pollination of mangoes, but more research is warranted. Heavy rainfall as a result of climate change can also affect cherry trees. The native capuli cherry [*Prunus serotina* subsp. *capuli* (Cav.) McVaugh], is a tree growing in the Andes in both Urban and wild conditions (Ramírez and Davenport 2016). Pollinators of the capuli cherry, such as bees and flies have been observed to stop floral visitation during heavy rains, considered out of season, that have been frequent during 2017. However, it needs to be determined how the heavy rainfall affects the pollination of the capuli cherry. Lately, on October 2, 2017 heavy rainfall coupled with a hail episode caused damage to urban trees at Bogotá, Colombia. Precipitation as hail within Bogotá city is an uncommon event, which can probably be attributed to climate change phenomena. The reader should keep in mind that Bogotá is located in the tropics and the annual mean temperature is about 13 °C and temperatures rarely fall below zero degrees. The 1–2 cm hail spheres impacted numerous tree species in some areas of the city, and caused leaf, floral and fruit drop, as well as leaf rupture in trees such as *Tecoma stans*, *Croton* sp. *Lafoesia acuminata*, *Syzygium paniculatum*, etc. (Fig. 3.2). Shrubs with large leaves e.g. Lulo (*Solanum quitoense*) also suffered leaf rupture by hail, which also caused some flowers and small fruits to drop (Ramírez Pers. Obs). Recently, fruit trees such as Feijoa (*Acca sellowiana*) have been negatively impacted by heavy rainfall coupled with hail events occuring towards the second semester of 2017 (August–December) in Bogotá, Colombia. Hail has caused floral, fruit drop and in some cases damaged leaves in 'Criollo' and 'Quimba' Feijoa trees (Ramírez and Kallarackal 2017, 2018). Climate change mitigation measures are required to protect or prevent hail from damaging trees. In the Amazon, changes in precipitation during the dry season caused by climate change conditions are considered a critical feature that will influence climatic fate of the region (Malhi et al. 2008).

Fig. 3.2 Impact of hail on trees in Bogotá, Colombia. (**a**) Tree with top leaves removed, (**b**) detail, (**c**) leaf abscission, (**d**) leaf damage in *Croton funkianus* and (**e**) leaves removed form *Tecoma stans* (Photos by Fernando Ramírez. Reproduced with permission)

Another severe impact of climate change is flooding. Heavy rains that occur as a consequence of climate change, through the dry period of the year, affect fruit trees within orchards that do not have an effective drainage system (Fischer et al. 2016). Waterlogged fruit trees devoid of adaptations to thrive in inundation, suffer from root anaerobic conditions that lead to anoxic conditions (Das 2012; Dwivedi and Dwivedi 2012). If the waterlogged conditions prevail, tree physiological events such as flowering and pollination are adversely impacted. The effects of waterlogging on pollination depend on the water level, whether the flowers are completely covered or not. Also, flowers could experience detachment or decay due to abnormal physiological conditions as a result of their internal decay process that begins with root anaerobic conditions (Das 2012; Dwivedi and Dwivedi 2012).

References

Bronstein JL, Hossaert-McKey MX (1995) Hurricane Andrew and a Florida fig pollination mutualism: resilience of an obligate interaction. Biotropica 27:373–381. https://doi.org/10.2307/2388922

Chapman CA, Chapman LJ, Struhsaker TT et al (2005) A long-term evaluation of fruiting phenology: importance of climate change. J Trop Ecol 21:31–45. https://doi.org/10.1017/S0266467404001993

Corlett RT, LaFrankie JV (1998) Potential impacts of climate change on tropical Asian forests through an influence on phenology. Clim Chang 39:439–453. https://doi.org/10.1023/a:1005328124567

Craig P, Trail P, Morrell TE (1994) The decline of fruit bats in American Samoa due to hurricanes and overhunting. Biol Conserv 69:261–266. https://doi.org/10.1016/0006-3207(94)90425-1

Crane JH, Campbell RJ, Balerdi CF et al (1993) Effect of Hurricane Andrew on tropical fruit trees. Proc Florida State Hortic Soc 106:139–144

Das H (2012) Agrometeorology in extreme events and natural disasters. BS Publikations, Hyderabad

Dwivedi P, Dwivedi R (2012) Physiology of abiotic stress in plants. Agrobios, Jodhpur

Fischer G, Ramírez F, Casierra-Posada F (2016) Ecophysiological aspects of fruit crops in the era of climate change. A review. Agron Colomb 34:190–199. https://doi.org/10.15446/agron.colomb.v34n2.56799

Kumar R (2014) Effect of climate change and climate variable conditions on litchi (*Litchi chinensis* Sonn.) productivity and quality. Acta Hortic 1029:145–154

Landry CL (2013) Changes in pollinator assemblages following hurricanes affect the mating system of *Laguncularia racemosa* (Combretaceae) in Florida, USA. J Trop Ecol 29:209–216. https://doi.org/10.1017/S0266467413000266

Malhi Y, Roberts JT, Betts RA et al (2008) Climate change, deforestation, and the fate of the Amazon. Science (80-) 319:169–172. https://doi.org/10.1126/science.1146961

Magrin GO, Marengo JA, Boulanger J-P et al (2014) Central and South America. In: Climate change 2014: impacts, adaptation and vulnerability: part B: regional aspects: Working Group II contribution to the fifth assessment report of the Intergovernmental Panel on Climate Change. Cambridge University Press, New York, pp 1499–1566

Negi VS, Maikhuri RK, Pharswan D et al (2017) Climate change impact in the western Himalaya: people's perception and adaptive strategies. J Mt Sci 14:403–416. https://doi.org/10.1007/s11629-015-3814-1

Potter KM, Crane BS, Hargrove WW (2017) A United States national prioritization framework for tree species vulnerability to climate change. New For 48:275–300. https://doi.org/10.1007/s11056-017-9569-5

Ramírez F, Davenport T (2012) Reproductive biology (physiology)—the case of mango. In: Valavi S, Rajmohan K, Govil J et al (eds) The mango. Studium Press, Houston, pp 56–81

Ramírez F, Davenport T (2016) The phenology of the capuli cherry [*Prunus serotina* subsp. *capuli* (Cav.) McVaugh] characterized by the BBCH scale, landmark stages and implications for urban forestry in Bogotá, Colombia. Urban For Urban Green 19:202–211

Ramírez F, Kallarackal J (2015) Responses of fruit trees to global climate change. SpringerBriefs. Springer, New York

Ramírez F, Kallarackal J (2017) Feijoa [*Acca sellowiana* (O. Berg) Burret] pollination: a review. Sci Hortic (Amsterdam) 226:333–341. https://doi.org/10.1016/J.SCIENTA.2017.08.054

Ramírez F, Kallarackal J (2018) Phenological growth stages of Feijoa [*Acca sellowiana* (O. Berg) Burret] according to the BBCH scale under tropical Andean conditions. Sci Hortic (Amsterdam) 232:184–190

Ramírez V, Ayala R, González H (2016) Temporal variation in native bee diversity in the tropical sub-deciduous forest of the Yucatan Peninsula, Mexico. Trop Conserv Sci 9:718–734

Rathcke BJ (2000) Hurricane causes resource and pollination limitation of fruit set in a bird-pollinated shrub. Ecology 81:1951–1958. https://doi.org/10.1890/0012-9658(2000)081[1951:HCRAPL]2.0.CO;2

Rosenzweig C, Phillips J, Goldberg R et al (1996) Potential impacts of climate change on citrus and potato production in the US. Agric Syst 52:455–479. https://doi.org/10.1016/0308-521X(95)00059-E

Scarano FR, Ceotto P (2015) Brazilian Atlantic forest: impact, vulnerability, and adaptation to climate change. Biodivers Conserv 24:2319–2331. https://doi.org/10.1007/s10531-015-0972-y

Snyder RL (2017) Climate change impacts on water use in horticulture. Horticulturae 3:27. https://doi.org/10.3390/horticulturae3020027

Chapter 4
Droughts and Pollination

Dry forests are environments characterized by drought conditions that extend for long periods with very few rainy periods. Under these circumstances trees have become adapted to survive under drought conditions (Fig. 4.1). Worldwide 42% of all intertropical vegetation and 49% of the vegetation of Mesoamerica (southern Mexico and Central America) and the Caribbean comprises tropical dry forest (Murphy 1995). Dry forest tree species distribution has been affected by climate change, e.g. southern Ecuador (Aguirre et al. 2017). To date, dry forests and forests worldwide face climate change impacts. These could cause several consequences within forests by varying the frequency, intensity, length, and timing of fire, drought, insect and pathogen outbreaks, invasive species, hurricanes, landslides, etc. (Dale et al. 2001).

The effect of drought on pollination has been established by several lines of evidence. Unusual droughts have a negative influence on trees not adapted to them. Drought causes reduced flowering, floral decay and delays or halts fruit development. It causes problems with flower pollination by decreasing the viability of pollen grains, reducing attractiveness of flowers to pollinators, and dimineshes nectar production within flowers (Alqudah et al. 2011). Also, insect pollinator activity can be delayed or completely hampered by drought events. Figs (*Ficus* spp.) and their specific wasp pollinators (Agaonidae) have been negatively impacted by a severe drought caused by El Niño-Southern Oscillation event. This event has increased the number of forest fires causing a seven-fold increase in adult fig tree mortality in contrast to non-drought years (Harrison 2000). The wasps are short lived and require a continuous supply of inflorescences for pollination to occur (Harrison 2000). The wasps complete their life cycle within the fig inflorescence. In Borneo, a severe drought caused a halt in the production of dioecious fig inflorescences causing a local extinction of fig wasps at the Lambir Hills National Park, Sarawak, Malaysia (Harrison 2000, 2001). No fig inflorescences occurred during a two-month period, which is twice the total life time span of wasps, preventing newly emerging wasps from finding inflorescences for breeding purposes (Harrison 2005). Whenever figs

© The Author(s) 2018   23
F. Ramírez, J. Kallarackal, *Tree Pollination Under Global Climate Change*,
SpringerBriefs in Agriculture, https://doi.org/10.1007/978-3-319-73969-4_4

Fig. 4.1 Dry forest environments at Curaçao (**a–b**) trees near Hato Cave, (**c**) Trees at the Shete Boka National Park which are near the ocean (**d**) and (**e**) *Acacia* sp. with thorns adapted to the dry environment. (Photos by Fernando Ramírez. Reproduced with permission)

are not pollinated by the specific wasps, fruit production is interrupted. This in turn, influences keystone species such as vertebrates that feed on the fruit and aid in seed dispersal (Harrison 2000), thus, causing major food-web problems within the forest.

Irregular drought in Asian forests have triggered mass flowering, increasing the opportunities for pollinators. The increased pollinator success is evident by greater fruit production and by the increased outcrossing and higher survival rates (Sakai et al. 2006). Similarly, warm and dry weather conditions during flowering, increased the activity and abundance of insect pollinators of trees (Vicens and Bosch 2000;

Willmer and Stone 2004). Olive, *Olea europaea* a wind-pollinated species is positively influenced by drier and warmer conditions, which improve pollen production and release (Galán et al. 2004; García-Mozo et al. 2008).

Drought, caused by climate change, is considered the greatest threat for Cocoa (*Theobroma cacao*) production in Africa (Schroth et al. 2016). This tree species is drought sensitive and West African cocoa yields have been impacted by severe El Niño drought years (Ruf and Schroth 2015). Drought could also impact cocoa pollinators, such as midges (Diptera, Ceratopogonidae) but more research is required. In India, within the Himachal Pradesh region, increasing temperature as a result of climate change has caused drought during summers and less snowfall during winters leaving the region unfit for apple cultivation (Singh 2013). This has forced farmers to shift to other crops and pollinators might have faced problems to adapt to the new conditions.

In Central America, fruit trees are more resistant to droughts in contrast to subsistence crops (rice, sorghum and beans), which are greatly impacted by lack of water (Imbach et al. 2017). Drought is a factor that negatively impacts the greatest amount of crops, closely followed by climate change driven non-seasonal precipitation and floods (Imbach et al. 2017). The effects of drought on pollination present a knowledge gap in which very little research has been conducted.

References

Aguirre N, Eguiguren P, Maita J et al (2017) Potential impacts to dry forest species distribution under two climate change scenarios in southern Ecuador. Neotrop Biodivers 3:18–29. https://doi.org/10.1080/23766808.2016.1258867

Alqudah AM, Samarah NH, Mullen RE (2011) Drought stress effect on crop pollination, seed set, yield and quality. In: Alternative farming systems, biotechnology, drought stress and ecological fertilisation. Springer, Dordrecht, pp 193–213

Dale VH, Joyce LA, Mcnulty S et al (2001) Climate change and forest disturbances. Bioscience 51:723–734. https://doi.org/10.1641/0006-3568(2001)051[0723:CCAFD]2.0.CO;2

Galán C, Vázquez L, García-Mozo H, Domínguez E (2004) Forecasting olive (*Olea europaea*) crop yield based on pollen emission. F Crop Res 86:43–51. https://doi.org/10.1016/S0378-4290(03)00170-9

García-Mozo H, Perez-Badía R, Galán C (2008) Aerobiological and meteorological factors' influence on olive (*Olea europaea* L.) crop yield in Castilla-La Mancha (Central Spain). Aerobiologia (Bologna) 24:13–18. https://doi.org/10.1007/s10453-007-9075-x

Harrison RD (2000) Repercussions of El Nino: drought causes extinction and the breakdown of mutualism in Borneo. Proc R Soc B Biol Sci 267:911–915. https://doi.org/10.1098/rspb.2000.1089

Harrison RD (2001) Drought and the consequences of El Niño in Borneo: a case study of figs. Popul Ecol 43:63–75. https://doi.org/10.1007/PL00012017

Harrison RD (2005) A severe drought in Lambir Hills national park. In: Pollination ecology and the rain forest canopy: Sarawak studies. Springer, New York, pp 51–64

Imbach P, Beardsley M, Bouroncle C et al (2017) Climate change, ecosystems and smallholder agriculture in Central America: an introduction to the special issue. Clim Chang 141:1–12

Murphy P (1995) Dry forest of Central America and the Caribbean. In: Bullock S, Mooney H, Medina E (eds) Seasonally dry tropical forest. Cambridge University Press, Cambridge, pp 9–34

Ruf F, Schroth G (2015) Economics and ecology of diversification: the case of tropical tree crops. Springer, Dordrecht

Sakai S, Harrison RD, Momose K et al (2006) Irregular droughts trigger mass flowering in aseasonal tropical forests in Asia. Am J Bot 93:1134–1139. https://doi.org/10.3732/ajb.93.8.1134

Schroth G, Läderach P, Martinez-Valle AI et al (2016) Vulnerability to climate change of cocoa in West Africa: patterns, opportunities and limits to adaptation. Sci Total Environ 556:231–241. https://doi.org/10.1016/j.scitotenv.2016.03.024

Singh IJ (2013) Impact of climate change on the apple economy of Himachal Pradesh: a case study of Kotgarh village. Ecology and Tourism 2013 (EcoTour-2013), 21–23 November 2013, Lviv, Ukraine

Vicens N, Bosch J (2000) Weather-dependent pollinator activity in an apple orchard, with special reference to *Osmia cornuta* and *Apis mellifera* (Hymenoptera: Megachilidae and Apidae). Environ Entomol 29:413–420. https://doi.org/10.1603/0046-225X-29.3.413

Willmer PG, Stone GN (2004) Behavioral, ecological, and physiological determinants of the activity patterns of bees. Adv Study Behav 34:347–466. https://doi.org/10.1016/S0065-3454(04)34009-X

Chapter 5
Plant-Insect Phenology and Pollination

Climate change has been known to impact plant pollination by changing flowering phenology and by distressing the activity of pollinators, e.g. flight (Abrol 2012). Similarly, phenological decoupling of plant–pollinator interactions (Settele et al. 2016) have been reported. Specifically, plants and insects have different responses to changing temperature, creating temporal (phenological) and spatial (distributional) disparities that cause problems at the population level (Reddy et al. 2013). Mismatches could impact plants by impairing decreased insect visitation that means less pollen deposition, whereas pollinators could face reduced food availability (Reddy et al. 2013). However, in some circumstances, pollinator–plant synchrony does not cause mismatches, due to generalist pollinator species keeping pace with changes in forage-plant flowering by switching between host plants (Fig. 5.1) (Settele et al. 2016). Animal biology and ecology associated with pollination i.e. population, reproductive aspects, and activity - flight, etc., are essential for understanding the impacts manifested by climate change. Relatively very little research has been conducted on the physiology of many crucial pollinators influenced by warming temperatures (Scaven and Rafferty 2013). This is evident in many tropical regions worldwide, where, animal pollinators comprise much more species and interactions, when compared to temperate conditions (Figs. 1.2 and 5.2).

Flower-insect synchronicity is an essential factor required in pollination, subsequent fertilization and fruit production in fruit trees. Reduced timing between flowering and the appearance of pollinators could decrease food sources for pollinators, leading to reduced pollinator abundance and increased extinctions of both pollinators and plants (Myers et al. 2017). Fruit tree-insect interactions within the context of climate change have been poorly examined in temperate, tropical and subtropical conditions (Figs. 5.3 and 5.4). Doi et al. (2008) reported that in temperate conditions in Japan, flowering in *Prunus* species (*Prunus davidiana*, P. × *yedoensis*, P. *pendula* form. *Ascendens*, and P. *armeniaca*) has occurred earlier in the past three decades and the appearance of a pollinator butterfly, *Pieris rapae* has been delayed (Doi et al. 2008). Flowering in *P. davidiana* occurred 12 days before the other

© The Author(s) 2018

F. Ramírez, J. Kallarackal, *Tree Pollination Under Global Climate Change*,
SpringerBriefs in Agriculture, https://doi.org/10.1007/978-3-319-73969-4_5

Fig. 5.1 Palm tree pollinators at the Dominican Republic (Photos by Fernando Ramírez. Reproduced with permission)

species. *P. yedoensis*, *P. armeniaca*, and *P. ascendens* currently initiate to flower prior to butterfly occurrence, but between 1950s and 1980s, the trees flowered 1 week after the first *Pieris rapae* individuals appeared (Doi et al. 2008). Besides, in 1998, flowering occurred in three *Prunus* species before the butterflies appeared. These tree-pollinator mismatches and early flowering are caused by warmer than usual days in the last three decades and even some cooler than usual days that can be attributed to climate change conditions (Doi et al. 2008). Tree-pollinator mismatches could have negative effects on the pollination and subsequent fruit set in *Prunus*. The rate of flowering advance and the advance appearance of its pollinators have been similar over a 46 year time period in New York State, USA (Bartomeus et al. 2013). The apple- pollinator relationship has been stabilized against climate change by a high diversity of floral visitors (Bartomeus et al. 2013). This relationship has advanced in both, apple and pollinators by a mean of 8 days since 1965. Bee species have shown phenological shifts considered either faster or slower than apple, therefore causing a stable phenological synchrony (Bartomeus et al. 2013).

Fig. 5.2 Bumblebees visiting the flowers of *Vallea stipularis* tree at La Calera, Cundinamarca State, Colombia (**a–d**). Bee visiting the flowers of *Oreopanax floribundum* at Bogota, Colombia (**e–f**) (Reproduced with permission)

Bee complementarity and diversity in apple pollination is a key aspect leading to phenological synchrony. Polce et al. (2014) examined climate-change mismatches between pears, apples, plums and other fruits and their pollinators under present and future climatic scenarios projected for 2050 for Great Britain. Under these scenarios, a geographical mismatch of fruit tree orchards and pollinators was detected under the featured climate change conditions. Projected scenarios for 2050 proposed that the most appropriate locations for orchards corresponded to low insect pollinator availability. This means low orchard pollination unless fruit tree production is transferred to more suitable climatic regions such as the north-westerly areas.

Fig. 5.3 Butterflies within a citrus orchard at Nocaima, Colombia (Top two photos). Numerous insects pollinate citrus in this particular environment. Butterflies pollinating the flowers of *Delonix regia*, a tropical tree at Melgar, Cundinamarca, Colombia (Bottom two photos). (Photos by Fernando Ramírez. Reproduced with permission)

Fig. 5.4 Citrus trees pollinated by honeybee (left) and fly (right) at Chipaque, Cundinamarca State, Colombia (Photos by Fernando Ramírez. Reproduced with permission)

However, wild pollinators' availability should be preserved in areas currently used for crop production, but more research is required to understand the implications for fruit tree pollination (Polce et al. 2014). Flowering in wind-pollinated plants has advanced more in contrast to insect-pollinated plants as a result of climate warming in Europe (Ziello et al. 2012; Gordo and Sanz 2009). In face of climate warming, wind-pollinated species might have changed their pollination mode as a response to unfavorable environmental conditions, allowing a better response to the variability of climate (Ziello et al. 2012). Temperate ecosystems have a greater frequency of wind-pollinated species, mainly trees (Kay and Sargent 2009). Conversely, in tropical forests, more than 90% of tree species are pollinated by biotic vectors (Bawa 1990; Jacobi and Carmo 2011). As a result of climate change, pollinator occurrence in fruit crops with great dependence on animal pollinators, e.g. avocado and guava are projected to decrease in future scenarios by 2050 in Brazil (Giannini et al. 2017). Furthermore, out of 13 crops, guava, coffee and mandarin will be the most affected by pollinator loss (Giannini et al. 2017).

Fruit tree- pollinator interactions have also been altered by climate change in the tropics. Particularly flight activity of pollinators (bees and flies) has been reduced by non-seasonal precipitation in mango orchards at La Mesa, Colombia (Ramírez pers. obs.). Moreover, the climate change effects on pollinators are associated with their thermal tolerance and plasticity to temperature changing conditions (Reddy et al. 2013). Mango trees produce less panicles and flowers due to altered phenology as a consequence of unusual rains resulting from climate change (Ramírez and Kallarackal 2015). This in turn, reduces floral resource availability for pollinators such as pollen and nectar. Heavy rains have also been linked to floral abscission in mango orchards in Thailand (Makhmale et al. 2016). Also, warmer than usual temperatures derived from climate change cause decreased pollinator activity in mango orchards (Bhriguvanshi 2010; Sajal et al. 2012). Heavy rain and hail have also impaired pollination mismatches between bumblebees and lulo (*Solanum quitoense*) flowers in Bogotá, Colombia. Observed pollinator visitation by bumblebees (*Bombus* sp.) has been delayed during the first week of October, 2017. Also, floral drop has been one of the consequences leading to reduced resources for pollinators (Ramírez Pers. Obs). This has also been observed in other tree species e.g. *Tecoma stans, Lafoesia acuminata, Prunus serotina* var. *capuli, Croton funkianus*, etc.

Among specialized tree-pollinator interactions, the wasp and fig interaction is considered a well studied relationship (Abrol 2012). Figs are considered keystone species in tropical communities because they provide food for a wide range of animals, namely, hornbills, toucans, parrots, pigeons, monkeys, bats and even fish (Abrol 2012). Most fig tree species are dependent on one or two fig wasp species (Agaonidae) (Abrol 2012; Jevanandam et al. 2013). An increase in temperature of 3 °C or more could decline the lifespan of four fig wasps, namely *Valisia malayana* (which pollinates *Ficus grossularioides*), *Ceratosolen appendiculatus* (pollinator of *F. variegata*), *Ceratosolen constrictus hewitti* (pollinator of *F. fistulosa*), and *Eupristina verticillata* (pollinator of *F. microcarpa*), all within the Agaonidae family (Jevanandam et al. 2013). This will reduce *Ficus* flower pollination / fig oviposition within flowers (Jevanandam et al. 2013).

Pollinator diversity and its interactions with both wild and native trees have also been disrupted by climate change events. In tropical Central America, sites with less forest cover and subjected to warmer – climate change driven temperatures loose more diversity, while locations with more forest cover and with less climate change warming retain diversity (Hannah et al. 2017). This influences the cultivation of coffee and tree crops that are dependent on native forest pollinators (Hannah et al. 2017). Thus, forested areas are valuable as pollinator reservoirs that could benefit crop production in adjacent agricultural systems (Hannah et al. 2017). Also, tropical forests can increase pollination services if the forested cover is augmented (Hannah et al. 2017).

The Intergovernmental Platform on Biodiversity and Ecosystem Services (IPBES) is a sound initiative focusing on pollinator status, values of pollination, and trends and drivers of change (IPBES 2016). This initiative has proposed a number of knowledge gaps and among them, the case of tropical developing nations of the world, where little is known about climate change and plant-pollinator interactions (IPBES 2016) has been mentioned. Recently, Ramírez and Kallarackal (2018) examined the tree-pollinator interactions within the context of IPBES, emphasizing the tropics and found that interactions between trees and pollinators in the tropics are complex, due to the fact that there are multiple pollinators and interactions e.g. arthropods, birds, etc. (Abrol 2012; Ramírez and Davenport 2013, 2016; Ramírez and Kallarackal 2017). Also, in the tropics, specialized tree pollinator interactions such as the fig-wasp symbiosis have shown mismatches (Jevanandam et al. 2013). But aside from this example, few investigations have examined the interactions between flowering time and pollinators. Other factors such as wind pollination in the tropics also require further research.

References

Abrol DP (2012) Pollination biology: biodiversity conservation and agricultural production. Springer, New York

Bartomeus I, Park MG, Gibbs J et al (2013) Biodiversity ensures plant-pollinator phenological synchrony against climate change. Ecol Lett 16:1331–1338. https://doi.org/10.1111/ele.12170

Bawa KS (1990) Plant-pollinator interactions in tropical rain forests. Annu Rev Ecol Syst 21:399–422. https://doi.org/10.1146/annurev.es.21.110190.002151

Bhriguvanshi S (2010) Impact of climate change on mango and tropical fruits. Westville Publishing House, New Delhi

Doi H, Gordo O, Katano I (2008) Heterogeneous intra-annual climatic changes drive different phenological responses at two trophic levels. Clim Res 36:181–190. https://doi.org/10.3354/cr00741

Giannini TC, Costa WF, Cordeiro GD et al (2017) Projected climate change threatens pollinators and crop production in Brazil. PLoS One 12:e0182274. https://doi.org/10.1371/journal.pone.0182274

Gordo O, Sanz JJ (2009) Long-term temporal changes of plant phenology in the Western Mediterranean. Glob Chang Biol 15:1930–1948. https://doi.org/10.1111/j.1365-2486.2009.01851.x

Hannah L, Steele M, Fung E et al (2017) Climate change influences on pollinator, forest, and farm interactions across a climate gradient. Clim Chang 141:63–75. https://doi.org/10.1007/s10584-016-1868-x

IPBES (2016) Summary for policymakers of the assessment report of the intergovernmental science-policy platform on biodiversity and ecosystem services on pollinators, pollination and food production. Secretariat of the Intergovernmental Science-Policy Platform on Biodiversity and Ecosystem Services. Bonn, Germany

Jacobi CM, Carmo do FF (2011) Life-forms, pollination and seed dispersal syndromes in plant communities on ironstone outcrops, SE Brazil. Acta Bot Brasilica 25:395–412. https://doi.org/10.1590/S0102-33062011000200016

Jevanandam N, Goh AGR, Corlett RT (2013) Climate warming and the potential extinction of fig wasps, the obligate pollinators of figs. Biol Lett 9:20130041. https://doi.org/10.1098/rsbl.2013.0041

Kay KM, Sargent RD (2009) The role of animal pollination in plant speciation: integrating ecology, geography, and genetics. Annu Rev Ecol Evol Syst 40:637–656. https://doi.org/10.1146/annurev.ecolsys.110308.120310

Makhmale S, Bhutada P, Yadav L, Yadav B (2016) Impact of climate change on phenology of mango – the case study. Ecol Environ Conserv 22:S127–S132

Myers SS, Smith MR, Guth S et al (2017) Climate change and global food systems: potential impacts on food security and undernutrition. Annu Rev Public Health 38:259–277. https://doi.org/10.1146/annurev-publhealth-031816-044356

Polce C, Garratt MP, Termansen M et al (2014) Climate-driven spatial mismatches between British orchards and their pollinators: increased risks of pollination deficits. Glob Chang Biol 20:2815–2828. https://doi.org/10.1111/gcb.12577

Ramírez F, Davenport TL (2013) Apple pollination: a review. Sci Hortic (Amsterdam) 162:188–203

Ramírez F, Davenport T (2016) The phenology of the capuli cherry [*Prunus serotina* subsp. *capuli* (Cav.) McVaugh] characterized by the BBCH scale, landmark stages and implications for urban forestry in Bogotá, Colombia. Urban For Urban Green 19:202–211

Ramírez F, Kallarackal J (2015) Responses of fruit trees to global climate change. SpringerBriefs. Springer, New York

Ramírez F, Kallarackal J (2017) Feijoa [*Acca sellowiana* (O. Berg) Burret] pollination: a review. Sci Hortic (Amsterdam) 226:333–341. https://doi.org/10.1016/J.SCIENTA.2017.08.054

Ramírez F, Kallarackal J (2018) Tree pollination and its conservation implications under global climate change conditions: IPBES and beyond. Wiley Clim Chang 9:e502. https://doi.org/10.1002/wcc.502

Reddy PVR, Verghese A, Sridhar V, Rajan VV (2013) Plant-pollinator interactions: a highly evolved synchrony at risk due to climate change. In: Climate-resilient horticulture: adaptation and mitigation strategies. Springer India, New Delhi, pp 295–302

Sajal R, Sthapit SR, Scherr SJ (2012) Tropical fruit tree species and climate change. In: Sthapit BR, Ramanatha Rao V, Sthapit SR (eds) Tropical fruit tree species and climate change. Bioversity International, New Delhi, pp 15–26

Scaven VL, Rafferty NE (2013) Physiological effects of climate warming on flowering plants and insect pollinators and potential consequences for their interactions. Curr Zool 59:418–426. https://doi.org/10.1093/czoolo/59.3.418

Settele J, Bishop J, Potts SG (2016) Climate change impacts on pollination. Nat Plants 2:16092. https://doi.org/10.1038/nplants.2016.92

Ziello C, Böck A, Estrella N et al (2012) First flowering of wind-pollinated species with the greatest phenological advances in Europe. Ecography (Cop) 35:1017–1023. https://doi.org/10.1111/j.1600-0587.2012.07607.x

Chapter 6
Conservation Implications

Perennial crop production has been reported to be sensitive to water availability, temperature, air pollution, solar radiation, CO_2, and drought (Alqudah et al. 2011; Glenn et al. 2013). These factors have been altered by today's climate change conditions. Under these conditions, tree phenology has been altered i.e. flowering has been delayed or advanced (Glenn et al. 2013; Ramírez and Kallarackal 2015, 2017). These aspects are considered a serious threat to the conservation of wild and cultivated tree species worldwide (Fig. 6.1).

Pollinators of tree species play an important role in reproduction by enabling fruit set. Currently, under the climate change, pollinators have been reported to decline. This situation has been analyzed by the Intergovernmental Platform on Biodiversity and Ecosystem Services (IBPES) which is an initiative of multiple research collaborators from around the world examining pollination aspects such as values, trends and pollinator status, and drivers of change (IPBES 2016; Ramírez and Kallarackal 2018). This initiative has established that the shift in abundance, ranges, and seasonal activities of some wild pollinator species have declined as a consequence of climate change (IPBES 2016). Also, it has established that climate change has caused plant-pollinator mismatches. Recently, Ramírez and Kallarackal (2018) examined IPBES gaps regarding plant-pollinator interactions, pollinator diversity and population attributes within the context of climate change, trees and conservation. Tree pollination services have been threatened by climate change, and particularly in the tropics more information is required (Ramírez and Kallarackal 2018). Furthermore, climate change effects i.e. elevated carbon dioxide, increased temperature, droughts, flooding, etc. have caused problems in plant-pollinator interactions (Ramírez and Kallarackal 2018). These gaps point out the need to conduct more research in the tropics and also, to establish conservation measures that act effectively to preserve tree and pollinator diversity. Scientists, governments and citizens from tropical countries are encouraged to provide possible solutions at the conservation level. This requires enacting conservation measures through suitable

© The Author(s) 2018   35
F. Ramírez, J. Kallarackal, *Tree Pollination Under Global Climate Change*,
SpringerBriefs in Agriculture, https://doi.org/10.1007/978-3-319-73969-4_6

Fig. 6.1 A cloud forest near the village of Chipaque, Cundinamarca State, Colombia (Photos by Fernando Ramírez. Reproduced with permission)

programs. However, to date few initiatives have been proposed to tackle this particular problem.

Conservation of trees and tree pollinators requires examination of the particular life history and reproductive strategies and knowledge on the effects of species-specific responses to elevated temperatures (Ramírez and Kallarackal 2018). In

Fig. 6.2 Conservation strategies to mitigate climate change (Photo by Fernando Ramírez. Reproduced with permission)

tropical environments, conservation measures to mitigate climate change impacts include introduction of trees to climatically suitable sites (Fig. 6.2) (Deb et al. 2017). In the case of Dipterocarp trees Sal (*Shorea robusta*) and Garjan (*Dipterocarpus turbinatus*) growing in tropical Asian forests, a strategy to mitigate the effects of climate change would be to introduce plantations into climatically suitable sites together with more care for trees within sites that have high risk in a future climate change scenario (Deb et al. 2017). These measures need further research in respect to the Dipterocarp pollinators, but provide a conservation strategy which is tree centered. According to Dawson et al. (2013) there are three ways to conserve tropical trees within smallholders' agroforests, (1) trees held or grown by farmers within agricultural landscapes where wild tree counterparts are to be found could be considered biodiversity reservoirs, (2) trees within farmland settings enable conservation by providing an alternative product source, which reduces the impact on wild trees, and by connecting to wild tree stands (Fig. 6.2). These conservation measures could be applied to tropical trees within the current climate change conditions. Moreover, it is required to investigate more about the pollinators of both agricultural and wild tree stands in the tropics. In the Amazon, there are opportunities for tree and forest conservation through the tropical forest carbon credits (Malhi et al. 2008). This system proposed monetary retribution to countries that reduce deforestation mainly in tropical regions (Santilli et al. 2005). It is sought to create a monetary large-scale incentive for reducing tropical deforestation and mitigate the effects of climate change by the conservation of forest stands within developing countries in the context of the Kyoto Protocol (Santilli et al. 2005). This system has been adopted in various countries worldwide (Wunder 2006; Pfaff et al. 2007; Neudert et al. 2017). In Costa Rica, carbon credits have helped to shape forest conservation and increased carbon sequestration (Pfaff et al. 2007). Aside from its benefits, several major problems arise while incorporating tropical forest carbon credits in developing countries. These problems include corrupt Governments, policies that are targeted to alleviate poverty might not effectively sequester carbon monetarily, in some developing countries compensation for payments for environmental services has been poorly tested, communities might be at risk from land-development by business type conservation, few are convinced about payment (supply side), carbon models could not be accurate at the site level, etc. (Wunder 2006; Pfaff et al. 2007; Preece et al. 2017). Tropical forest carbon credits focus is on forest stand conservation and could also protect tree pollinators, but more data and knowledge is required. Commonly the carbon projects have been targeted at the tree level and have not included pollinators.

Numerous conservation attempts have been conducted in respect to temperate and subtropical tree conservation. Adaptation strategies according to Howden et al. (2007), Foley et al. (2011), Iglesias et al. (2012), NCA (2012), Glenn et al. (2013), Ravi and Mustaffa (2013), Rajan et al. (2013), Reddy (2015) and Singh (2013) include changes in farm management at the local level, enhanced water management, using plant genetic resources within the context of sustainability, raising awareness among farming communities, developing suitable technologies, changing cultivars and altering planting dates, developing uniform phenological monitoring for fruit trees e.g. mango, etc. (Fig. 6.2). In the case of subtropical crops such as litchi production in India, conservation strategies include (1) windbreaks for avoiding damage, (2) improved rootstocks, (3) management of the canopy, (4) girdling, (5) replacing new for old trees, (6) mulching and (7) employing honey bees as potential pollinators (Kumar and Nath 2013). Regarding tropical plantation crops, e.g. coconut, rubber, oil palm, cashew, and cocoa, Hebbar et al. (2013) reported that approaches to mitigate climate change include the adoption of more tolerant tree varieties, and inclusion of the best management practices. In the case of coconut, mitigation strategies devise the use of scientific technologies, e.g. drip irrigation, implementation of drought tolerant cultivars, soil and water conservation practices (Hebbar et al. 2013). In India, climate change is a threat for cashew cultivation. Drought mitigation strategies for this crop include mulching, water and soil conservation, and conservation agriculture, drip irrigation and fertigation particularly during fruit development (Rupa et al. 2013). In the United States of America, the national prioritization framework for tree species vulnerability to climate change established a categorization based on risk factors (1) climate change exposure, (2) sensitivity to climate change, and (3) adapting capacity to climate change (Deb et al. 2017). This assessment has established conservation measures for current and future climate change scenarios (Fig. 6.1) (Deb et al. 2017). These investigators also included pollination vectors either wind, insects or mammals which provide a more consistent risk assessment. According to Dawson et al. (2011) trees can respond to climate change by (1) the facilitated movement of environmentally-suitable germplasm to other proper geographic scales, (2) increasing tree stands population sizes by farm management e.g. pollinators and (3) The use of more 'flexible' species and populations. According to Matocha et al. (2012) climate change mitigation and adaptation can be assessed simultaneously because of the financial and technical advantages. First, planning is required for mitigation under current and future climate change scenarios and form mitigation investments, second, several land-use interventions could bear both benefits for adaptation and mitigation and third, carbon-based finances could provide funding for adaptation. Within these initiatives it will also be useful to consider tree pollinators that could lead to a better understanding for adaptation and mitigation. Another strategy to counteract climate change is through the identification of tree genetic resources that could adapt to climate change conditions (Fig. 6.2) (Abrol 2012). Genetic variation as part of a tree's genetic pool could allow adaptation within agricultural systems e.g. promising new varieties could replace old ones (Glenn et al. 2013). Also, specific features and/or genes could be introduced within existing cultivars through genetic engineering (Fig. 6.2) (Hajjar and Hodgkin 2007). However, these methodologies are controversial because they drastically modify the plant genetics and also could cause negative impacts on the environment. Moreover, other conservation strategies are available to establish a diverse pollinator assemblage that bears different features and responses to ambient conditions (Abrol 2012).

Pollinator conservation strategies under climate change conditions are important aspects that need to be considered. Tree pollinators include a great number of arthropod species, mammals, bats, birds, etc. Major pollinator groups comprise bees, beetles, wasps, butterflies, and flies (Reddy et al. 2013). Conservation strategies for pollinators require understanding of physiological aspects e.g. responses to higher temperatures, reproduction, dispersion, feeding, flight related aspects, etc. Conservation management efforts depend on species specific traits and geographic range whether tropical, subtropical or temperate. Issues pertaining to the importance of pollinator diversity conservation focus on three factors, namely, practical, theoretical, or abstract (Ollerton 2017). In the case of butterfly conservation, Coristine et al. (2016) reported that improved climate change-related dispersal should be based on improving landscape connectivity according to species-specific richness, mobility, and climate change, as well as landscape permeability. Moreover, effective conservation strategies elsewhere to mitigate climate change include landscape areas that enable pollinators to move to restored or managed locations for them (Ollerton 2017). Another conservation measure is to establish a high floral visitor diversity within plant–pollinator phenological synchrony to mitigate climate change (Bartomeus et al. 2013).This has been effective in apple orchards, where different bee species show phenological changes that are slower or faster than the flowering period of apple trees, enabling effective pollinator and floral coupling (Bartomeus et al. 2013). Other strategies to mitigate climate change include farming with alternative pollinators (Christmann and Aw-Hassan 2012). In Central America adaptation strategies of pollinator services include (1) reservoir areas that enhance bee diversity or act as source areas for recolonization after climate change driven changes i.e. droughts, and (2) restoration areas with enhanced forest canopy cover which could aid to improve the pollinator services (Hannah et al. 2017). Mitigation for the impacts of climate change on bumblebees includes nesting site and forage availability or a combination of both through habitat management (Williams and

Osborne 2009). Several countries in Europe and North America have management and stewardship that seek to improve landscapes (Williams and Osborne 2009). This can be done through forest conservation and maintaining heterogenous landscapes for agriculture, which in the case of coffee, requires live fences, shade trees, as well as native plants that protect, provide food and nesting sites-materials (Imbach et al. 2017). Furthermore, climate change requires reviewing insect management strategies and goals in view of a new kind of scientific engagement through management decision-making (Hellmann et al. 2016). These refer to interdisciplinary research, use of process-based modeling, and increased participation of entomologists within social decision making (Hellmann et al. 2016).

References

Abrol DP (2012) Pollination biology: biodiversity conservation and agricultural production. Springer, New York

Alqudah AM, Samarah NH, Mullen RE (2011) Drought stress effect on crop pollination, seed set, yield and quality. In: Alternative farming systems, biotechnology drought stress and ecological fertilisation. Springer, Dordrecht, pp 193–213

Bartomeus I, Park MG, Gibbs J et al (2013) Biodiversity ensures plant–pollinator phenological synchrony against climate change. Ecol Lett 16:1331. https://doi.org/10.1111/ele.12170

Christmann S, Aw-Hassan A (2012) Farming with alternative pollinators (FAP)—an overlooked win-win-strategy for climate change adaptation. Agric Ecosyst Environ 161:161–164. https://doi.org/10.1016/J.AGEE.2012.07.030

Coristine LE, Soroye P, Soares RN et al (2016) Dispersal limitation, climate change, and practical tools for butterfly conservation in intensively used landscapes. Nat Areas J 36:440–452. https://doi.org/10.3375/043.036.0410

Dawson IK, Vinceti B, Weber JC et al (2011) Climate change and tree genetic resource management: maintaining and enhancing the productivity and value of smallholder tropical agroforestry landscapes. A review. Agrofor Syst 81:67–78. https://doi.org/10.1007/s10457-010-9302-2

Dawson IK, Guariguata MR, Loo J et al (2013) What is the relevance of smallholders' agroforestry systems for conserving tropical tree species and genetic diversity in circa situm, in situ and ex situ settings? A review. Biodivers Conserv 22:301–324. https://doi.org/10.1007/s10531-012-0429-5

Deb JC, Phinn S, Butt N, McAlpine CA (2017) The impact of climate change on the distribution of two threatened Dipterocarp trees. Ecol Evol 7:2238–2248. https://doi.org/10.1002/ece3.2846

Foley JA, Ramankutty N, Brauman KA et al (2011) Solutions for a cultivated planet. Nature 478:337–342. https://doi.org/10.1038/nature10452

Glenn D, Kim S, Ramirez-Villegas J, Läderach P (2013) Response of perennial horticultural crops to climate change. In: Janick J (ed) Horticultural reviews, vol 41. Wiley, Hoboken, pp 47–130

Hajjar R, Hodgkin T (2007) The use of wild relatives in crop improvement: a survey of developments over the last 20 years. Euphytica 156:1–13. https://doi.org/10.1007/s10681-007-9363-0

Hannah L, Steele M, Fung E et al (2017) Climate change influences on pollinator, forest, and farm interactions across a climate gradient. Clim Chang 141:63–75. https://doi.org/10.1007/s10584-016-1868-x

Hebbar KB, Balasimha D, Thomas GV (2013) Plantation crops response to climate change: coconut perspective. In: Climate-resilient horticulture: adaptation and mitigation strategies. Springer India, New Delhi, pp 177–187

Hellmann J, Grundel R, Hoving C, Schuurman GW (2016) A call to insect scientists: challenges and opportunities of managing insect communities under climate change. Curr Opin Insect Sci 17:92–97. https://doi.org/10.1016/J.COIS.2016.08.005

Howden SM, Soussana J-F, Tubiello FN et al (2007) Adapting agriculture to climate change. Proc Natl Acad Sci U S A 104:19691–19696. https://doi.org/10.1073/pnas.0701890104

Iglesias A, Garrote L, Quiroga S, Moneo M (2012) A regional comparison of the effects of climate change on agricultural crops in Europe. Clim Chang 112:29–46. https://doi.org/10.1007/s10584-011-0338-8

Imbach P, Fung E, Hannah L et al (2017) Coupling of pollination services and coffee suitability under climate change. Proc Natl Acad Sci U S A 114:10438–10442. https://doi.org/10.1073/pnas.1617940114

IPBES (2016) Summary for policymakers of the assessment report of the Intergovernmental Science-Policy Platform on Biodiversity and Ecosystem Services on pollinators, pollination and food production. Secretariat of the Intergovernmental Science-Policy Platform on Biodiversity and Ecosystem Services. Bonn, Germany

Kumar R, Nath V (2013) Climate resilient adaptation strategies for litchi production. In: Climate-resilient horticulture: adaptation and mitigation strategies. Springer India, New Delhi, pp 81–88

Malhi Y, Roberts JT, Betts RA et al (2008) Climate change, deforestation, and the fate of the Amazon. Science 319:169–172. https://doi.org/10.1126/science.1146961

Matocha J, Schroth G, Hills T, Hole D (2012) Integrating climate change adaptation and mitigation through agroforestry and ecosystem conservation. Springer, Dordrecht, pp 105–126

NCA (2012) National climate assessment. Climate change and agriculture: effects and adaptation. USDA. U.S. Government Printing Office, Washington, DC

Neudert R, Ganzhorn J, Wätzold F (2017) Global benefits and local costs – the dilemma of tropical forest conservation: a review of the situation in Madagascar. Environ Conserv 44:82–96. https://doi.org/10.1017/S0376892916000552

Ollerton J (2017) Pollinator diversity: distribution, ecological function, and conservation. Annu Rev Ecol Evol Syst 48:353–376. https://doi.org/10.1146/annurev-ecolsys-110316

Pfaff A, Kerr S, Lipper L et al (2007) Will buying tropical forest carbon benefit the poor? Evidence from Costa Rica. Land Use Policy 24:600–610. https://doi.org/10.1016/j.landusepol.2006.01.003

Preece ND, van Oosterzee P, Hidrobo Unda GC, Lawes MJ (2017) National carbon model not sensitive to species, families and site characteristics in a young tropical reforestation project. For Ecol Manag 392:115–124. https://doi.org/10.1016/j.foreco.2017.02.052

Rajan S, Ravishankar H, Tiwari D et al (2013) Harmonious phenological data: a basic need for understanding the impact of climate change on mango. In: Climate-resilient horticulture: adaptation and mitigation strategies. Springer India, New Delhi, pp 53–65

Ramírez F, Kallarackal J (2015) Responses of fruit trees to global climate change. SpringerBriefs. Springer, New York

Ramírez F, Kallarackal J (2018) Tree pollination and its conservation implications under global climate change conditions: IPBES and beyond. Wiley Clim Chang 9:e502. https://doi.org/10.1002/wcc.502

Ravi I, Mustaffa MM (2013) Impact, adaptation and mitigation strategies for climate resilient banana production. In: Climate-resilient horticulture: adaptation and mitigation strategies. Springer India, New Delhi, pp 45–52

Reddy PP (2015) Climate change adaptation. In: Climate resilient agriculture for ensuring food security. Springer India, New Delhi, pp 223–272

Reddy PVR, Verghese A, Sridhar V, Rajan VV (2013) Plant-pollinator interactions: a highly evolved synchrony at risk due to climate change. In: Climate-resilient horticulture: adaptation and mitigation strategies. Springer India, New Delhi, pp 295–302

Rupa TR, Rejani R, Bhat MG (2013) Impact of climate change on cashew and adaptation strategies. In: Climate-resilient horticulture: adaptation and mitigation strategies. Springer India India, New Delhi, pp 189–198

Santilli A, Moutinho P, Schwartzman S et al (2005) Tropical deforestation and the Kyoto protocol an editorial essay. Clim Change 71:267–276. https://doi.org/10.1007/s10584-005-8074-6

Singh HCP (2013) Adaptation and mitigation strategies for climate-resilient horticulture. In: Climate-resilient horticulture: adaptation and mitigation strategies. Springer India, New Delhi, pp 1–12

Williams PH, Osborne JL (2009) Bumblebee vulnerability and conservation world-wide. Apidologie 40:367–387. https://doi.org/10.1051/apido/2009025

Wunder S (2006) The efficiency of payments for environmental services in tropical conservation. Conserv Biol 21:48–58. https://doi.org/10.1111/j.1523-1739.2006.00559.x

Chapter 7
Conclusion

Climate change poses a severe threat to the pollination of numerous tree species worldwide. While the effect of climate change has been evaluated in a number of cases in herbaceous plants and their pollinators, less is known about the interactions between trees and pollination and especially climate change extreme events such as flooding, droughts, temperature increase and changes in precipitation particularly in trees growing in tropical regions. Thus, more research is required in wild as well as in cultivated trees in tropical environments. The influence of both elevated temperature and cool temperatures imposed by climate change conditions on pollen physiology of trees, e.g. viability, growth and pollen tube germination are detrimental for the pollination and in some cases hampers fertilization and subsequent fruit set. Most studies have focused on the effect of elevated temperature within economically important tree crops, but there is a need for more research on wild tree pollens, which would provide scientists with a bigger picture to understand the effects climate change conditions. It is also important to focus on the optimal temperatures for pollen germination in less investigated crop species and among well investigated cultivated trees. Tree pollen seasons have been modified by climate change especially by increased temperatures. Numerous investigations have been carried mainly in Europe and North America describing pollen season advancement or delay over several years. However, less is known about this topic in the tropics. It is suggested that phenological events in tropical regions such as pollen season and flowering need to be described and quantified to better understand the effects of climate change impacts. The effects of elevated and cool temperatures also need to be determined for tree pollinators. Very little is known about the effect of temperature on flight, foraging and feeding activity of tree pollinators in tropical environments. Other climate change events such as unusual precipitation need more investigation under tropical environments in relation to pollinator activity. Climate change extreme events such as hurricanes have provided insight about their devastating effect on trees and their pollinators. Hurricanes can completely wipe out a whole pollinator population and then, subsequent recolonization can be achieved by the

F. Ramírez, J. Kallarackal, *Tree Pollination Under Global Climate Change*, SpringerBriefs in Agriculture, https://doi.org/10.1007/978-3-319-73969-4_7

pollinator dispersal abilities. Recent hurricane events provide unique opportunities for studying pollinator and tree interactions which can provide meticulous information about natural restoration and mitigation after a severe event. Precipitation as rain, hail, and snow has also been modified by climate change conditions. Precipitation often halts pollinator activity and causes devastating effect when falling as hail often causing leaf, flower and fruit abscission and damaging pollinator hives and habitats. This is particularly the case of tropical regions, where precipitation as hail, an unusual event has generated impacts on trees, but its short term and long term effects require more research. Water logging is another climate change impact which has been evaluated in trees, and it is known to hamper pollination if the water level rises above the tree's reproductive structures. Also, if roots are submerged for prolonged periods without proper aeration, foliage, flowers and fruits are likely to suffer abortion. Climate change imposed drought has caused reproductive issues in trees including flower decay, abscission and in specialized trees like figs it has caused pollinator declines or removal from a particular area. Specialized fruit tree species such as figs are keystone species because numerous vertebrates, such as mammals, birds as well as the whole food-web are dependent upon them. This raises the question about the importance of pollinators for both cultivated and wild environments which occur through plant-pollinator interactions, in this book, tree-pollinator interactions. These interactions have been thoughtfully studied in herbaceous plants in temperate conditions. Less is known about woody angiosperms and their interactions with pollinators particularly in the tropics where thousands of tree species are awaiting to be studied. Three pollinator mismatches have been evaluated in temperate conditions where pollinators e.g. butterflies have been documented to occur later than the flowering event. In the tropics the interplay of numerous pollinators, i.e. increased pollinator diversity could hamper the effects of climate change factors such as precipitation, decreased or increased temperatures, etc. But more research is indeed warranted to fully understand pollinator diversity and tree pollination in tropical regions worldwide. Describing and quantifying the effects of climate change on pollinators and trees is an important step to establish management and conservation measures for mitigating and adapting to climate change conditions. Effective conservation initiatives are required to preserve both wild and cultivated trees and their pollinators. These measures include farm management, water management, incorporating new technologies, using different tree cultivars and genetic resources, or establishing frameworks for tree conservation under climate change conditions. All of these initiatives solve the problem at least partly and could probably lead to provide consistent solutions. Other conservation strategies have focused on pollinators, identifying ways to enhance landscape features, establish pollinator reservoir areas, generating heterogeneous agricultural landscapes, etc. In spite of all conservation measures to mitigate and adapt to climate change conditions, climate change phenomenon needs to be tackled at its source empowered by consumerism, and also industrial, political and economic criteria in today's globalized world. The change is more about the attitude of we humans towards nature. It is a change in philosophy from a materialistic view, purely based on generation of wealth regardless of the environmental consequences to a higher philosophical view where the natural world is connected to humans. Pollination in the light of climate change is an essential aspect that has motivated numerous researchers worldwide to provide their view and research initiative. It is important to generate awareness about pollination studies worldwide and particularly under climate change conditions. Developing countries require programs to evaluate and determine the effects of climate change, pollinators and trees. This book is sought to contribute by providing the personal view of the authors and by a literature review of the main aspects that pertain to trees and their pollinators.